Climate Change and Water Resources:

A Primer for Municipal Water Providers

Awwa
Research
Foundation

About the Awwa Research Foundation

The Awwa Research Foundation (AwwaRF) is a member-supported, international, nonprofit organization that sponsors research to enable water utilities, public health agencies, and other professionals to provide safe and affordable drinking water to consumers.

The Foundation's mission is to advance the science of water to improve the quality of life. To achieve this mission, the Foundation sponsors studies on all aspects of drinking water, including supply and resources, treatment, monitoring and analysis, distribution, management, and health effects. Funding for research is provided primarily by subscription payments from approximately 1,000 utilities, consulting firms, and manufacturers in North America and abroad. Additional funding comes from collaborative partnerships with other national and international organizations, allowing for resources to be leveraged, expertise to be shared, and broad-based knowledge to be developed and disseminated. Government funding serves as a third source of research dollars.

From its headquarters in Denver, Colorado, the Foundation's staff directs and supports the efforts of more than 800 volunteers who serve on the board of trustees and various committees. These volunteers represent many facets of the water industry, and contribute their expertise to select and monitor research studies that benefit the entire drinking water community.

The results of research are disseminated through a number of channels, including reports, the Web site, conferences, and periodicals.

For subscribers, the Foundation serves as a cooperative program in which water suppliers unite to pool their resources. By applying Foundation research findings, these water suppliers can save substantial costs and stay on the leading edge of drinking water science and technology. Since its inception, AwwaRF has supplied the water community with more than $300 million in applied research.

More information about the Foundation and how to become a subscriber is available on the Web at **www.awwarf.org**.

Climate Change and Water Resources:

A Primer for Municipal Water Providers

Prepared by:
Kathleen Miller and David Yates
National Center for Atmospheric Research*
3450 Mitchell Lane, Boulder, CO 80301

With assistance from:
Conrad Roesch and D. Jan Stewart
National Center for Atmospheric Research*
3450 Mitchell Lane, Boulder, CO 80301

*The National Center for Atmospheric Research (NCAR) is sponsored by the National Science Foundation and managed by the University Corporation for Atmospheric Research

Jointly sponsored by:
Awwa Research Foundation
6666 West Quincy Avenue, Denver, CO 80235-3098
and

University Corporation for Atmospheric Research (UCAR)
PO Box 3000, Boulder, CO 80307

Published by:

DISCLAIMER

Table of Contents

**Climate Change Information in Utility Planning
and Adaptive Management**...55

Conclusion 69

Glossary...71

Index 75

References 79

Acknowledgments

Many individuals contributed information, insights and assistance to this project. Participants in the March, 2004 *Workshop on Climate Change and Water Utilities*, held at the National Center for Atmospheric Research, provided valuable input on the organization and content of this document. In particular, the authors would like to thank the following individuals and their respective organizations for assisting with the development of case study materials: Alan Chinn – Seattle Public Utilities; Carol Ellinghouse – City of Boulder; Paul Fesko – City of Calgary Waterworks; Bertha Goldenberg – Miami-Dade Water and Sewer Department; Lorraine Janus – New York City Department of Environmental Protection; Joan Kersnar – Seattle Public Utilities; Mark Knudson – City of Portland Water Bureau; Rick Langley – Greenville Utilities Commission; John Loughry – Denver Water; Arthur Meuleman – Kiwa Water Research; Peter Spillett – Thames Water Utilities; Gary Tilkian – Metropolitan Water District of Southern California; Marc Waage – Denver Water; Rocky Wiley – Denver Water. We wish to extend special thanks for scientific input and advice to: Michael Dettinger – US Geological Survey and Scripps Institution of Oceanography; Rob Wilby – Environment Agency, UK; Connie Woodhouse – Paleoclimatology Branch, NOAA National Climatic Data Center, and to the following NCAR Scientists: Aiguo Dai, Jerry Mahlman, Kevin Trenberth, and Tom Wigley. Valuable insights on management and assessment issues were provided by: Alvin Bautista – Los Angeles Department of Water and Power; Alex T. Bielak – National Water Research Institute, Environment Canada; Bonnie Colby – Department of Agricultural Economics, University of Arizona; Alan Hamlet – JISAO/ SMA Climate Impacts Group, University of Washington; Roger Jones – RS Climate Risk and Integrated Assessment, CSIRO Atmospheric Research; David C. Major – Center for Climate Systems Research, Earth Institute, Columbia University; and Robert C. Wilkinson – Environmental Studies, University of California, Santa Barbara. In addition, we wish to express sincere gratitude to Conrad Roesch for valuable research assistance, to D. Jan Stewart for editorial assistance, and to Anne Oman and Jean Renz for workshop support.

Synopsis:
Climate Change and Water Utilities

There is a great deal of misunderstanding surrounding the subject of climate change, often leading to profound confusion regarding its potential impacts on natural resource systems and human wellbeing. Well-intentioned, but misguided attempts by the popular press and movie industry to call attention to the prospect of climate change have left much of the public with the impression that the Earth's climate system is either poised at the brink of cataclysmic change or that global climate change is a myth that they can safely ignore. Neither of those extreme views provides useful guidance to anyone attempting to make informed decisions about the management of climate-sensitive resources.

Here, we will attempt to dispel some of the confusion by summarizing the best available scientific evidence on climate change – including both natural changes and changes that may be caused by human activities. In particular, this Primer will focus on what is known about the implications of climate change for the water cycle and the availability and quality of water resources. The goals of this primer are to 1) introduce water utility managers to the science of climate change; 2) suggest the types of impacts it can have on water resources; and 3) provide guidance on planning and adaptation strategies. This guidance primarily reflects the activities of forward-looking utilities that have begun to plan and prepare for these changes, with some additional insights gained from the research community.

> *To plan efficiently, it is important to understand how and why climate may change in the future and how that may affect the resources upon which the water utility industry depends.*

Water industry professionals are keenly aware of the fact that climate variability affects the availability and quality of water resources and that runoff or temperature extremes can affect their operations. Unanticipated extremes, such as an unprecedented drought, are likely to pose particularly severe problems. Prudent management focuses on anticipating and mitigating the potential adverse impacts of such natural variability. To plan efficiently, it is important to understand how and why climate may change in the future and how that may affect the resources upon which the water utility industry depends.

Will climate change have significant impacts in the near future on water availability, water quality and the ability of water utilities to meet the needs of their customers at desired levels of reliability and affordability? If so, what types of impacts could occur? What should utilities be doing to assess and prepare for the resulting risks and opportunities? Is this an issue that requires attention now, or will climate change occur so far in the future that water utilities can safely ignore it and concentrate on more pressing concerns? These are the types of questions addressed in the following pages.

• *The Earth's climate is changing and will continue to change in coming decades*

Temperature records from around the world show a significant warming trend over large areas of the Earth's surface since the beginning of the 20th century, with a rapid acceleration of warming in recent decades. In short, climate change is already happening. Over the past century, global average surface temperature increased by approximately 0.6° C (Figure 1).

There is strong scientific evidence that human activities have contributed to this warming, for example, by increasing the concentration of greenhouse gases in the atmosphere. Burning fossil fuels releases carbon dioxide into the atmosphere. Concentrations of this important greenhouse gas are rising rapidly. This suggests that rising global surface temperatures and associated climate changes will continue, and likely accelerate over the next several decades and beyond.

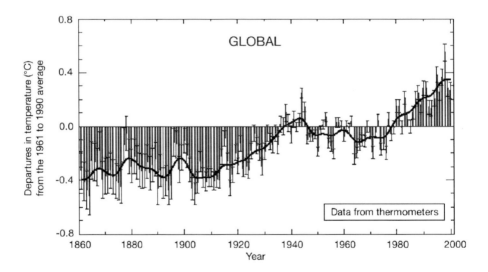

Figure 1. Variations of the Earth's surface temperature for the past 140 years (1860 to 2000). Source: IPCC WGI 2001, p. 26.

- *Warmer temperatures have already had significant hydrologic impacts*

Climate changes consistent with the impacts of global warming are occurring now. During the past half century, warmer temperatures have resulted in significant changes in the seasonal timing of runoff in many mountainous areas. In the western U.S and southwestern Canada, spring snowpacks have been smaller and have been melting earlier in most mountain areas (Figure 2). These declines have often occurred despite increases in total winter precipitation in those locations. Earlier spring melting and reduced spring snowpacks have been especially evident in the Cascade and northern Sierra Nevada Mountains, where winter temperatures are relatively mild. Some higher elevation mountain locations in the Southern Sierra Nevada and Rocky Mountain ranges have shown an increasing trend in April 1 snowpacks, but even there the peak in spring runoff is generally occurring earlier (Stewart et al. 2004).

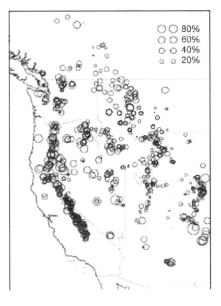

Figure 2. Observed linear trends (1950-1997) in April 1 snow water equivalent (SWE) relative to the starting value for the linear fit at 824 snow course locations in the western U.S. and Canada, with negative trends shown by red circles and positive trends by blue circles. (Source: Mote 2004, p. 2.)

- *Water utilities are vulnerable to climate-related disruptions even in the absence of climate change*

Climatic extremes often pose challenges for water utilities. For example, heavy runoff events frequently result in deterioration of source water quality, necessitating additional treatment costs, and increasing the risk of contamination of water supplies by disease pathogens. Floods also can threaten supply system infrastructure. In 1999, for example, Greenville, North Carolina received heavy precipitation from two hurricanes that struck in quick succession. The second of the two, Hurricane Floyd, left the city's water treatment plant surrounded by floodwaters and resulted in damages to the utility's infrastructure, costing approximately $11 million (Figure 3). Only timely sandbagging kept the water sufficiently at bay to allow the facility to continue operating. Following

Figure 3. Greenville Utilities water treatment plant surrounded by floodwaters. September, 1999 (Photo courtesy of Greenville Utilities Commission).

that event, the utility has built a protective berm around the facility to protect it from future floods.

Droughts can rapidly decrease a utility's ability to service all water demands. Severe droughts in recent years have led many water utilities to impose emergency restrictions on their customers. For example, several dry years preceded a severe drought in 2002, which left many Colorado reservoirs dangerously low (Figure 4). As a consequence, urban water customers were limited to very short periods of outdoor water use.

Figure 4. Denver's Dillon Reservoir during June 2002 (Photo courtesy of Denver Water).

- *Drought-related wildfires can increase a utility's vulnerability to water quality impairment and loss of reservoir storage capacity as a result of soil erosion and debris flows.*

There have been a number of large forest fires in western North America over the past several years. Such fires are likely to become more common as warmer temperatures and earlier loss of snowpacks lead to increased vegetation stress and reduced summer soil moisture. Fires can have serious impacts on downstream water quality and reservoir sedimentation. For example, a flash flood event following the Buffalo Creek fire in 1996 resulted in severe sediment and debris flows into Denver's Strontia Springs Reservoir. The immediate damage control and cleanup required during that episode, together with longer-term water quality impacts have imposed heavy costs on the utility (Figure 5).

Such climate-related extremes may become both more common and more difficult to anticipate in the future because of global climate change. To learn more about why that is so, and what forward-looking water utilities are doing to adapt to the changes, read on. The Primer begins with brief case studies that illustrate how water utilities are already considering climate change in their planning processes. Next, we provide an overview of the science of climate change and the evidence supporting projections of future climate change. Later sections examine the potential impacts on the hydrologic cycle and on water resources. We draw upon further case studies to examine the implications of such changes for water utilities and to discuss the adaptive strategies that utilities might employ.

Figure 5. Debris flow into Strontia Springs Reservoir on July 12, 1996 as a result of the Buffalo Creek fire and flash flood (Photo courtesy of Denver Water).

Introduction:
Issues and Perspectives

The scientific evidence for human-caused global climate change has become quite compelling in recent years. Global average temperatures have been rising, and human activities have changed the composition of the atmosphere significantly enough that we can now confidently say that the climate will continue to change. Along with the projected future warming, there will be changes in atmospheric and oceanic circulation, and in the hydrologic cycle, leading to altered patterns of precipitation and runoff. There also will likely be impacts on other physical and biological systems. For example, a warmer climate will make sea level rise inevitable. There also will be new stresses on ecological systems, including forests and riparian zones as well as coastal and freshwater aquatic systems. Such stresses may affect the regulatory environment for water utilities. Water provision for environmental needs is an important policy consideration in many locations, and climate change may make it more difficult to achieve a satisfactory balance between human water uses and environmental stewardship.

Scientists agree on some of the important broad-scale features of the expected hydrologic changes, the most likely of which will be an increase in global average precipitation and evaporation as a direct consequence of warmer temperatures. That, however, does not mean that there will be more precipitation everywhere or that runoff and recharge would increase in proportion to precipitation.

At the regional scale, precipitation predictions are less certain. Changes in circulation patterns will be critically important in determining future changes in precipitation and water availability, and climate models can provide only a crude picture of how those patterns may change. The currently available evidence suggests that arctic and equatorial regions may have a tendency to become wetter, and that subtropical regions may experience drying. Projections of precipitation changes for temperate regions are less consistent.

...scientific theory suggests an intensification of the global hydrological cycle...

The water supply for any utility will depend on the quantity and timing of local and regional precipitation, both of which may change with global climate change. While it is impossible to make reliable predictions of changes in the overall quantity of precipitation for a particular region, scientific theory suggests an intensification of the global hydrological cycle, leading to more intense but possibly less frequent periods of precipitation.

In other words, we may see longer periods of drought alternating with spells of heavy rainfall and runoff. Such changes could create a number of difficulties for water utility planning and operations. For example, greater runoff variability could make it more difficult to maintain optimal reservoir levels, which could reduce the reliability of water storage. In addition, increased reliance on groundwater resources during extended dry spells could reduce aquifer levels and discharge to surface water bodies, which could cause unintended damage to freshwater ecosystems.

The direct effects of temperature changes on water supplies are also significant, particularly for the timing of runoff. For example, in mountainous regions, there will likely be shorter snow accumulation periods – especially in lower elevation areas, possibly leading to reduced annual snowpacks, earlier spring melting and reduced late summer flows. Warmer temperatures during the winter will affect the form of precipitation, with a larger fraction of total precipitation coming as rain rather than snow. However, when it does snow, warmer temperatures and increased moisture availability may result in heavier snowfalls. A temperature change of only a few degrees during the melting season would have a substantial effect on the timing of spring runoff. Less snowpack in the late spring means that there will be a smaller supply in late summer, when water is scarcest and demand is high.

Unfortunately, at the regional and local scales that are relevant for water utilities, current scientific understanding does not yet allow confident projections of the magnitude or precise nature of climatic changes. Therefore, important uncertainties remain regarding how regional and local climates, hydrology and ecosystems will change in the coming century. Because regional and local variables are what matter for municipal water management, the prospect of climate change has imposed a new level of uncertainty on water managers. This suggests that they will need to examine and adapt their methods of planning to account for the fact that past hydrological patterns may become an increasingly unreliable guide to the future.

Why is climate change of particular interest to water resource planners? First, given the nature of the industry, decisions made in the near term will affect system reliability well into the future. For instance, utilities build costly water infrastructure with the expectation that investments will meet future requirements for decades to come. Likewise, policies designed to improve the efficiency of water use take time to implement, and it takes a long time to achieve their full benefits. Second, long-term planning in the context of uncertainty is already standard practice in this industry. Water utilities must account for many future uncertainties when formulating long-term plans, such as potential changes in water consumption patterns due to demographic and socio-economic changes. Climate change is an additional source of uncertainty that will become increasingly relevant to water resource managers in the 21st century. Just as with any other source of uncertainty, best practice requires understanding as much as possible about the changes that can occur and their implications for operation and management of the utility.

The following case studies introduce some of the hydrological changes that may result from climate change and demonstrate how pioneering utilities have already incorporated these possibilities into future planning. These case studies introduce a theme that will be apparent throughout the Primer: planning for uncertainty. Each example illustrates general problems that climate change could pose, but also highlights the fact that effective planning requires understanding that any future climate scenario involves uncertainty. Thus, in all cases, these utilities have developed precautionary strategies. That is, their decisions have focused on making their systems and operations sufficiently robust, resilient and flexible to meet future needs, given a broad range of possible changes in water supply and demand. For example, Seattle water managers are taking a cue from the American Association of State Climatologists (AASC), which argues that policy responses to climate variability and change should be flexible and sensible. Competing views of the long-term climate future may be equally credible given the difficulty of prediction and the impossibility of verifying predictions decades into the future. The AASC takes the position that policies related to long-term climate not be based on particular predictions, but instead should focus on policy alternatives that make sense for a wide range of plausible climatic conditions regardless of future climate. Climate is always changing on a variety of timescales, and being prepared for the consequences of this variability is a wise policy.

Snowmelt and Seasonal Runoff: Supply Infrastructure Planning in Portland, OR and Seattle, WA

As will be seen in the section on the science of climate change, one of the most confident predictions concerning global climate change is that average temperatures over land masses will increase. Warmer temperatures during the winter will affect the form of precipitation, with more of the precipitation falling as rain rather than snow. This may lead to increased streamflow during winter, earlier peak runoff in spring and reduced late summer streamflow. An average temperature change of only a few degrees during the melt season would very likely alter the timing of spring runoff.

In the Pacific Northwest, both the Portland Water Bureau and Seattle Public Utilities have begun to examine how such changes in seasonal runoff could affect their ability to meet customer demands, particularly during the summer months, as well as

> *...one of the most confident predictions concerning global climate change is that average temperatures over land masses will increase. This may lead to increased streamflow during winter, earlier peak runoff in spring and reduced late summer streamflow. An average temperature change of only a few degrees during the melt season would very likely alter the timing of spring runoff.*

9

instream flows to support fish habitat. Their efforts can provide lessons for many other cities in the western United States as well as many in Europe, Northern Asia and Canada who face similar potential impacts on system reliability.

Portland and Seattle are alike in that both utilities rely heavily on surface water coming from protected watersheds where runoff timing depends partly on snowmelt. Both utilities are located in a region characterized by large seasonal differences in precipitation. Most of the precipitation occurs during the winter, and summers are relatively dry. The utilities differ, however, in the options available for supply enhancement. The direction and emphasis of their planning and management efforts reflect those differences. The Portland Water Bureau has explored various climate change scenarios to examine the robustness of its supply system. Based on this examination, utility managers are developing adaptive strategies to cope with the possible changes in seasonal flows, and other likely effects of climate change.

The Portland Water Bureau supplies water to approximately 800,000 people in the Portland metropolitan area, with deliveries totaling about 40 billion gallons per year. Its primary water source is the Bull Run Watershed, an unfiltered surface supply that has two reservoirs with a combined usable storage capacity of 10 billion gallons. The utility also has a backup groundwater supply located along the south shore of the Columbia River. Precipitation over the watershed typically ranges between 59 to over 80 inches per year, with most falling during the winter months. Therefore, supplying water during the summer months is the greatest challenge for this utility. Summer water demand can peak at over 220 million gallons a day, which is double the average daily use. Climate change could increase these difficulties in coming decades, but the Portland Water Bureau has already begun to prepare for such an eventuality by considering changes in long-term water demand and supply in their current infrastructure planning.

To evaluate the implications of future warming, the Portland Water Bureau conducted an analysis based on future climate scenarios derived from four different climate models (Palmer and Hahn 2002). They ran the climate scenarios through watershed hydrology, regional population growth, and system management models to simulate the impacts on system reliability and reservoir conditions. While the specific results vary, general trends such as increased winter precipitation, earlier snowmelt, and drier summers were consistent across the models. Such scenarios suggest that in the absence of additional supply infrastructure, future climate change scenarios may lead to decreased reliability of supply with a concurrent increase in summer water demand, leading to an increase in overall system vulnerability. Already, the utility is preparing for an increase in seasonal demand of between 8 and 10 percent, primarily resulting from population growth. This is equivalent to an additional billion gallons of water demand during the summer, which is about 10 percent of the current storage capacity of existing surface reservoirs. Under current climate conditions, it would be possible to meet this additional demand by fully exploiting the summer groundwater supply, but using groundwater as a primary source rather than a backup would remove a safety net for drought years and thereby decrease system reliability.

Current scientific evidence suggests that the primary threat that climate change poses for Portland's water supply is not a reduction in annual average precipitation but rather a change in runoff timing that aggravates a deficiency in storage capacity. The Portland Water Bureau's analysis suggests that changes in runoff timing could result in winter flows increasing by as much as 15 percent, accompanied by a 30 percent decrease in late spring flows. The impact of such changes, together with an increase in summer demand, would result in a 2.8–5.4 billon gallon decrease in reservoir storage volume by the end of the drawdown period or about 15 to 30 percent of the current storage. The utility is considering these factors as it evaluates the feasibility of further expanding the existing groundwater supply and/or expansion of existing source-water reservoirs. The proposed projects will increase system reliability under both current conditions and various climate change scenarios, and therefore constitute a precautionary strategy. Greater surface storage will also decrease the frequency of groundwater use, and therefore increase the sustainability of that emergency supply.

In addition to supply augmentation, the Portland Water Bureau is evaluating other strategic measures, such as conjunctive use strategies that coordinate the optimal use of existing surface and groundwater supplies, including use of aquifer storage and recovery (ASR). Since the exact nature of climate change is unknown, Portland is also emphasizing flexibility in infrastructure development. For example, as an unfiltered surface water supply, Portland may be especially vulnerable to extreme storm events in winter months that would result in elevated turbidity, making the surface water supply temporarily unusable. While Portland anticipates remaining unfiltered for the foreseeable future, water treatment options have been considered that might readily accommodate the addition of filtration equipment in the future.

Seattle Public Utilities is actively working with the University of Washington's Joint Institute for the Study of the Atmosphere and Oceans (JISAO) Climate Impacts Group. This is a group of scientists and policy analysts at the university that is examining the potential impacts of climate change within the Pacific Northwest region (which is defined as the states of Washington, Oregon and Idaho, and the Columbia River Basin). The UW JISAO Climate Impacts Group believes that by planning now, and by incorporating information about climate variability and change into decisions about how the region manages its natural resources, resource managers and decision-makers can reduce the negative impacts of (and take better advantage of the opportunities brought by) both human-caused climate change and natural variability.

Seattle Public Utilities supplies water to 1.3 million people and businesses in the Seattle metropolitan region. Its sources provide approximately 50 billion gallons per year to its retail and wholesale customers. Nearly all this water is from the 90,000-acre Cedar River Watershed and the 13,300-acre South Fork Tolt River Watershed located on the western slopes of the Cascade Mountains in eastern King County, WA. Water use is approximately the same as it was in the mid-1960s, despite population growth, due to conservation savings and system efficiencies. Recent climate statistics inform us that the Seattle area receives over 75 percent of its annual precipitation from the beginning of October through the end of March. Snowpack accumulations in the mountain watersheds normally peak around the beginning of April and runoff from snowmelt usually ends by July. Peak snowpack accumulations for the Cedar and Tolt watersheds above elevation 2500 feet average around 30 inches of snow water equivalence. The summer months are typically mild with minimal amounts of precipitation. The average annual precipitation for the Seattle area is about 37 inches; Cedar Lake at elevation 1560 feet receives about 100 inches, and the Tolt weather station at elevation 2000 feet receives about 90 inches. Average annual air temperature in Seattle is about 52°F; Cedar Lake is about 47°F, and Tolt weather station is about 44°F.

Systematic monitoring and collection of meteorological and hydrological data such as precipitation, air temperature and streamflow began as early as the late 1800s in the central Puget Sound region, and agencies such as the US Geological Survey and the National Weather Service maintain historical records. Traditionally, water planners assume that these historical records, when they are long enough, represent sufficient variability to support analysis of present and future water system performance and behavior. Indeed, water projects have been built and instream flow requirements have been established in the region based primarily on available historical weather and streamflow data.

Under the current hydrological regime, the storage reservoirs on both Seattle sources are refilled in the spring by snowmelt and rainfall. Water stored in the form of snowpack is an important element in managing reservoir refill because, once it has accumulated, it becomes a known quantity that can be relied on to refill the reservoirs. On the other hand, spring rainfall events are uncertain and difficult to predict. This uncertainty is significant in balancing the need for full reservoirs at the start of summer water use with the need for storage capacity to regulate downstream flows during steelhead spawning and incubation periods. Seattle Public Utilities is well aware of the sensitivity of its system to changing snowpack, and routinely monitors the condition of the snowpack when making decisions regarding reservoir operations. The utility has been managing reservoir levels using a "dynamic rule curve" that adjusts reservoir refill targets based on real-time snowpack and soil moisture conditions. This approach uses actual conditions to adjust reservoir management and increase the likelihood of full reservoir refill prior to the summer reservoir drawdown period.

Seattle Public Utilities has also started to examine a set of climate change scenarios. Like Portland, they are now sponsoring climate change research work by the UW JISAO Climate Impacts Group to explore and develop analysis techniques that will enable regional water planners and decision-makers to incorporate global climate change information into local long-range water supply planning processes.

Water planners at Seattle Public Utilities believe it is important to recognize that translating climate change scenarios down to regional and local scales with confidence is a significant problem for researchers and practitioners. One of the management challenges that a climate change study presents is that observed temperature changes over the past 100 years are not uniform throughout the world. Some places show trends of greater warming, while other locations are actually getting cooler, despite the fact that on average the Earth as a whole has been getting warmer. Similarly, there are probably regional and local differences. For example, we may find that impacts to the Central Puget Sound region may be different than those projected for the Columbia River Basin and eastern Washington, and there may even be differences between local watersheds within the Central Puget Sound region. Historic surface air temperature data for mid-elevation in the Cedar River watershed, for example, show a cooling trend expressed over the last 70 years.

The research work that Seattle Public Utilities is sponsoring by the UW JISAO Climate Impacts Group will first examine the current state-of-the-art climate change prediction models used by the Intergovernmental Panel on Climate Change (IPCC). This examination will focus on the levels of uncertainty associated with climate change scenarios generated by these climate prediction models, and the steps necessary to prepare climate change model results for use in existing regional or watershed-level models (a.k.a. downscaling techniques). The research effort will then be poised to develop useful analysis techniques using computer simulation modeling methods to evaluate potential future climate change impacts to local water resources. The research team will give special attention to identifying and documenting the important uncertainties and complexities associated with the methods studied to give water planners an understanding of the limitations of the method as well as to help identify focus areas where the method can be improved in future efforts. The water planner can then keep these uncertainties in mind when expensive or long-term water projects are considered. This research effort is underway and is expected to be completed in 2005.

Glacier Recession and Water Conservation Strategies: Calgary, Alberta

The City of Calgary, Alberta, Canada withdraws its water from two rivers, the Bow and a major tributary to the Bow River, the Elbow. Both rivers originate in the Rocky Mountains to the west of the city. Both rivers receive substantial contributions to flow from snowpack and glacier melt; however, the Bow River, about 10 times the size of the Elbow, can have a substantial contribution from glacier melt in dry years. Like snowpack, the volume and melt timing of glaciers are sensitive to temperature change. Scientists

Scientists have documented glacier recession across the globe, and this trend will likely continue if temperatures continue to rise.

have documented glacier recession across the globe, and this trend will likely continue if temperatures continue to rise. For a period of time, glacier-dominated systems may see an increase in runoff because of rapid melting, but in the long run less water will be stored in the glacier. As in the case of snowpack decline, it is unclear if average annual runoff will increase or decline, but there is a strong probability that in the long run glacial runoff will contribute less to river flows during the summer season than it does at present.

The Bow and Elbow Rivers are the sources of supply for the City of Calgary Waterworks' service population of one million. Calgary does not have significant reservoir storage. The Glenmore Dam and Reservoir on the Elbow River is relatively small with a storage capacity of approximately 20 days at current rates of consumption. System reliability depends to a large extent upon consistent river flows.

The larger Bow River receives a substantial, although highly variable, portion of its summer flow from several glaciers. Two glaciers, the Bow and Crowfoot that feed the Bow River, have receded substantially over the past several decades. On average, summer flow contribution from glacier melt ranges from 4.6 to 7.5 percent (Table 1). During a dry year, glacier melt can contribute as much as 47.4 percent of the August flow in the Bow River (Table 1), so the future condition of the glacier will have a substantial impact on Calgary's water.

Table 1
Contribution of glacier melt to the Bow River (Courtesy of Calgary Waterworks)

Month	Average for 1970 to 1998	1970 – Low Flow Year
July	4.6%	28.3%
August	7.5%	47.4%
September	5.0%	35.1%

Calgary's utilities have collaborated with scientists to anticipate the future of glacier-fed flows. Their findings predict that average annual flows will be sufficient, but continued glacier recession will result in lower flows in August and September, when stress on water supply is greatest (Hopkinson and Young 1998). In addition to climate change, Calgary Waterworks faces a rapidly increasing customer base, as the current population growth rate of Calgary is between 2.5 and 3 percent per year.

In the past, Calgary has not had a problem with insufficient supply. In fact, the utility has never implemented water restrictions because of supply scarcity. Historically, water conservation has not been a major issue, so there is a large capacity to improve the efficiency of water use. Calgary Waterworks has taken advantage of this opportunity as a key part of the utility's strategy to prepare for climate change and other stresses on water supply. In coming years, the utility plans to be capable of supplying an additional half million customers, a fifty percent increase in current population, with the current volume of water withdrawal. One approach to reach this goal is to reduce per capita water use by up to one-third by encouraging conservation. Another strategy involves increasing the efficiency of the water treatment process. Already treatment plant upgrades have begun which will ultimately allow the utility to fully utilize the water that it withdraws from the river by full recycling of the filter to waste and backwash water created by the treatment process. This will eliminate the utility's discharge of water back to the river from its water treatment plant.

Decreasing per capita water use will largely depend on the cooperation of the utility's customers and other departments of the City of Calgary, so the utility has developed a public education program that includes publications for customers and school programs on conservation issues. Conservation incentives have also received heavy attention, including programs offering rebates for water-saving appliances. Water metering is also an important demand-side tool to give consumers a monetary incentive to conserve. Calgary is currently migrating from flat rate to metered accounts, so that every home will have a meter by 2014. The case for metering is well known and, in general, comparison of customer classes in Calgary has shown that metered customers on average use 50 percent less water than flat rate customers.

The public sector of the city of Calgary is also implementing water conservation strategies. For example, the city has recently upgraded about 2000 irrigation systems for public areas to include meters. Many of these are already connected to a centralized weather-linked control system, and the plan is to eventually link all city-owned irrigation systems to central control. In addition to the direct savings in water expenditures, systems such as these are important for publicity. If the utility expects individuals to conserve water or to cooperate if the city must impose water restrictions, then it is important that citizens know the city is also committed to prudent water use. While water restrictions have not been a significant issue in the past, Calgary has realized they may soon be needed. To prepare for such a possibility, bylaws that outline how water restrictions are to be implemented in the event of a severe drought have been reconsidered and updated.

Calgary Waterworks has emphasized that public outreach will play an essential role in climate change adaptation strategies. When approaching water scarcity problems from the demand side, public cooperation is clearly vital for successful conservation. In addition, supply side initiatives, such as infrastructure investment, will also need support (and funding) from the utility's customers. It will be much easier to gain that support from a public that is aware of the possibility of climate change and the problems it poses for the water utility.

Climate Variability, Demand Management and Infrastructure Projects – England and Wales (UK)

The last introductory case study focuses on water utilities in England and Wales, and their experiences with planning for climate change in the context of recurring drought and flood episodes. The government privatized the major water utilities in 1989 and organized them according to the boundaries of the former Water Authority units. The duties of these privatized utilities are defined by statute and include cost-effective provision of clean water, treatment of wastewater, and environmental responsibilities. These broadly defined duties, and the fact that the utilities are organized on the basis of catchment basins, makes integrated water resource assessment and management both easier and more necessary for these utilities than may be the case for utilities in other settings. That, in turn, has facilitated the process of assessing climate change impacts and response options.

Water utilities in England and Wales face heavy regulation with respect to both pricing and performance of their statutory duties. Regulation involves a regular cycle of planning that is subject to intense public scrutiny. The government now encourages a "twin-track" approach to supply planning that simultaneously considers both resource development and demand management options. The government also has actively encouraged utilities to consider climate change when formulating their long-range water resource plans. In doing so, the utilities work closely with regulatory agencies such as the Environment Agency. One of the methods that they use to deal with uncertainties in demand and supply projections is to calculate a "headroom" factor that would be needed to meet supply security targets. The headroom calculations incorporate allowances for the possible impacts of climate change.

Assessment efforts have also been helped by the coordination of industry and government-funded research efforts through UKWIR (UK Water Industry Research), which is a water industry research group funded by the utilities. The UK government has invested a considerable amount of money in climate change research and supports a number of research centers across the country. The water utility industry has formed successful research partnerships with these centers to evaluate the risks that climate change may pose for the utilities.

Longer dry spells are one of the likely consequences of global climate change. In fact, droughts have played a significant role in shaping changes in water planning and management in England during the past quarter-century, and in focusing the attention of UK water utilities on planning for climate change (Subak 2000). There have been four major droughts during this period: the first occurred in 1976 through 1977, the second in 1984, followed by a three-year drought beginning in 1989, and finally a single-year drought in 1995.

The 1995 drought prompted the national government to mandate that all major water supply companies begin to examine climate change scenarios when calculating demand and supply balances. At the request of the Department of the Environment, English and Welsh utilities began to prepare detailed plans for adapting to global climate change following the 1995 drought. They have based these planning efforts on global climate change scenarios derived from various climate model projections and on climatic trends that utilities have observed in recent years. The model projections tend to suggest wetter winter conditions and hotter summers, with an increased likelihood of extended dry spells. Recent climate observations vary from one region to the next, but many of England's utilities reported drier summers or more intense but less frequent periods of precipitation in comparison to the historical record. While these scenarios do not provide a clear picture of the likely magnitude of future changes in hydrology, the utilities are using them to explore and assess the types of problems that climate change could pose.

An interesting aspect of the recent drought experiences is that despite similar hydrological characteristics, each drought evoked different responses by water resource managers. Expansion of water storage capacity was the main response to the 1976–1977 drought. While the resulting infrastructure projects did not solve all problems during later droughts, they greatly increased the ability of many utilities to respond to the subsequent drought events. After 1977, five of England's ten major water regions built new reservoirs, and the utility managers in these districts observed that they would not have been capable of supplying sufficient water during the 1995 drought without these infrastructure investments. The droughts during the 1980s also resulted in several supply infrastructure projects, but demand management began to play a role in some regions with the introduction of water use restrictions. The drought episodes demonstrated that dealing with peaks in demand presented the biggest challenges for most of England's utilities. Such demand spikes are intensified and therefore most obvious during droughts.

Responses to the 1995 drought focused much more on demand management than in previous years. Since 1995, utilities have worked in earnest to implement a number of projects to reduce consumption, including metering water and reducing system leakage. Other infrastructure development has followed these droughts, with most construction projects focused on developing redistribution capacity.

In the context of future planning, utilities are considering adaptation strategies that they have applied following past droughts. Some utilities, for example Essex and Suffolk, are vulnerable to increased runoff variability because their storage volume is insufficient to capture enough runoff during high flow periods to supply longer dry periods. As a result, these utilities have developed

...utility managers must consider how their systems are sensitive to climate change and identify what options are available to reduce these risks.

plans for reservoir addition or expansion. The Thames Water Utility is considering storage expansion to prepare for future summer droughts, but also has invested in regional distribution changes to cope with peak demands. In addition to enhanced regional distribution systems, several regions are planning to increase transfers from one region to another; for example, Severn Trent negotiated arrangements to import water from the Anglian region.

Wetter winters and episodes of intense precipitation are some of the possible consequences of climate change that climate model projections suggest for the UK. In that context, UKWIR has funded research on the implications of heavy rainfall events for the design of sewerage systems. The potential water quality impacts of inadequate sewerage design in the event of heavy runoff is a significant concern for the industry. Utilities have dealt with the uncertainties associated with projections of wetter winters and hotter summers by exploring the merits of various options for storing winter runoff for later use under a wide range of climate scenarios. In summary, forward-looking planning for climate change has become a standard practice for water utilities in England and Wales.

The previous case studies demonstrate that utilities may have different vulnerabilities to climate change and could differ in their adaptation options. There is no single best strategy to prepare for climate change. Rather, individual utility managers must consider how their systems are sensitive to climate change and identify what options are available to reduce these risks. The following chapters on climate change and hydrology provide background information to guide water utility managers in thinking about the impacts of climate change on the water resources on which they rely. Additional case studies describe how some utilities are weighing the likelihood of specific hydrological changes and gauging the risks they pose to system reliability. This Primer concludes with a further discussion of adaptive strategies, illustrated with examples drawn from utilities in a variety of settings.

The Science of Climate Change

What is climate and what does "climate change" mean?

We all know that weather varies from day to day, that it changes with the seasons, and that no two years are ever exactly alike. One way to describe the distinction between weather and climate is that "climate is what you expect, and weather is what you get." In other words, weather describes the evolution of the current state of the atmosphere, while climate is a measure of the typical weather for a particular place, hour of day, and time of year. Climate, as a statistical concept, measures not only expected average conditions, but also the characteristic range of variability of those conditions. Climate change will alter the likelihood of various types of weather events.

The climate system, as shown in Figure 6, includes the atmosphere, oceans, ice, land, vegetation, and surface water. The Sun and human activities act as important external influences on the climate system. Interactions among all of these components determine the geographical and seasonal distribution of climates across the surface of the globe. A change in any of these elements will cause changes in global and regional climates. The process may involve a sequence of adjustments and feedbacks in other components of the system. For example, there is a positive feedback between temperature and ice cover. Suppose that a period of increased solar input causes air temperatures to warm. That would tend to reduce the total area covered by ice and snow. Because snow and ice are

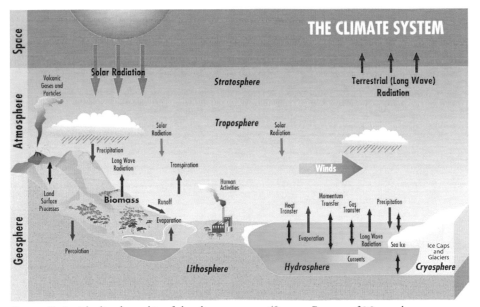

Figure 6. An idealized graphic of the climate system (Source: Bureau of Meteorology 1992. The Greenhouse Effect and Climate Change, Bureau of Meteorology, copyright Commonwealth of Australia reproduced by permission.).

very bright, they reflect sunlight back into space. With less ice and snow, the surface of the planet would be darker and would absorb more solar radiation. That, in turn, would cause further warming in the affected areas.

The Sun is the source of energy that drives the climate system. Solar radiation heats the atmosphere and the surface of the Earth. To balance the amount of energy coming in from the Sun, the Earth must radiate the same amount of energy back to space – in the form of infrared radiation. Greenhouse gases, which include water vapor, carbon dioxide, methane, nitrous oxide and a variety of human-made chemical compounds, trap some of the outgoing infrared radiation. When the energy balance is upset, for example by increases in the amount of greenhouse gases in the atmosphere, then the Earth will warm until a new balance is established, centuries later.

Figure 7 provides a globally averaged view of the Earth's energy budget. The term "greenhouse effect" refers to the fact that the atmosphere absorbs most of the infrared radiation leaving the surface of the Earth and re-emits part of that energy back toward the earth's surface. Increased concentrations of greenhouse gases in the atmosphere act to increase the "back radiation" term on the right-hand side of the figure. That, in turn, would warm the Earth's surface. Increased loss of energy through infrared radiation, and the release of latent heat through increased evapotranspiration and precipitation are among the processes that act to restore the energy balance.

The global water cycle plays an important role in the global energy balance because evaporation and cloud formation help to regulate both incoming and outgoing radiation. Water vapor is itself one of the most important greenhouse gases, and because a warmer atmosphere can hold more water vapor, it provides a powerful positive

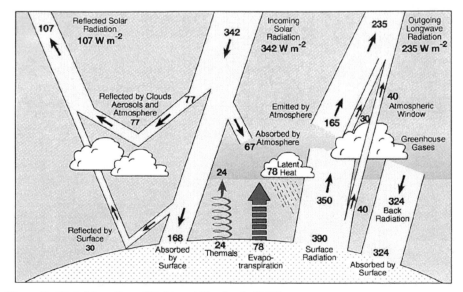

Figure 7. Global heat flows (Kiehl and Trenberth 1997).

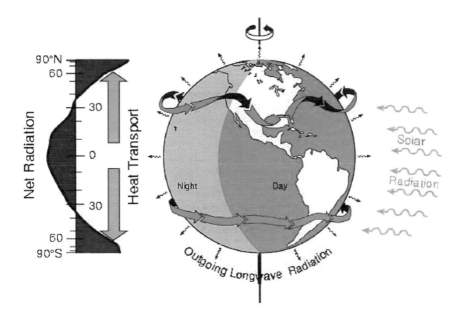

Figure 8. Heating dynamics of the Earth (courtesy of Kevin Trenberth, after Trenberth et al.

feedback to other sources of warming. Clouds, in particular, play a complicated role in the energy balance. They act as a blanket – warming the Earth's surface by absorbing and emitting thermal radiation. On the other hand, they also act to cool the surface of the Earth by reflecting incoming sunlight back into space. These opposing effects almost cancel each other out, but in our current climate, clouds appear to have a slight net cooling effect.

Because the Earth is a sphere, the Sun's heating is uneven (Figure 8). There is an energy surplus near the equator and a deficit near the poles. The circulation of the atmosphere and oceans transports heat from the tropics toward the poles, making the Earth's tropical regions cooler, and its polar regions warmer, than they would be if the Earth had no atmosphere or oceans.

The atmosphere and oceans are constantly in motion. That

> *Water vapor is itself one of the most important greenhouse gases, and because a warmer atmosphere can hold more water vapor, it provides a powerful positive feedback to other sources of warming.*

motion displays some stable patterns, which define contrasting climatic zones. For example, the Intertropical Convergence Zone (ITCZ) is a broad band that girdles the equator, characterized by rising air, frequent convective storms, and high annual precipitation. Just north and south of the ITCZ, centered at latitudes around 30 degrees north and south, are bands of hot, dry, descending air that create deserts in the world's subtropical regions. In the temperate regions, storms are steered by broad wind bands, called jet streams, that flow from west to east. The position of each jet stream migrates with the changing seasons, and planetary waves of high- and low-pressure regions develop within the jet streams.

These vary over time, but they are anchored, to some extent, on underlying geographical features such as mountains and boundaries between oceans and land. That anchoring results in semi-permanent predominant storm tracks that help to define the characteristics of regional climates.

Figure 9 is a generalized picture of these circulation patterns.

During each hemisphere's winter season, there is a greater imbalance between the energy deficit at that pole and the energy surplus in the tropics. This contributes to the formation of storm fronts, as the poleward transport of heat intensifies.

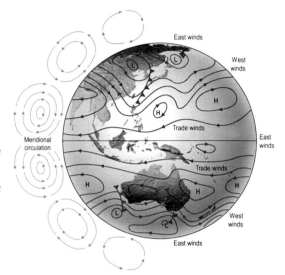

Figure 9. Generalized representation of circulation patterns (WMO 2003, p. 46).

In addition to these broad global climate patterns, the nature of local climates depends on such things as proximity to large bodies of water and the location of mountain ranges. For example, the windward side of a mountain range generally receives considerably more precipitation than nearby locations on the downwind side.

What one calls a climate change depends on the time period being considered. Climate varies naturally from one year to the next, and over decades and centuries as well, so the distinction between climate variability and climate change is somewhat fuzzy. Any trend or persistent change in the statistical distribution of climate variables (temperature, precipitation, humidity, wind speed, etc.) constitutes a climate change. Regional climate changes may result from persistent changes in the details of oceanic and atmospheric circulation. For example, the El Niño-Southern Oscillation (ENSO) phenomenon causes changes in the distribution of heat within the Pacific Ocean and the surrounding atmosphere. That, in turn, leads to changes in predominant storm tracks. The effects on local climates can be striking, with some areas receiving much heavier than normal precipitation, while other areas experience severe drought.

Figure 10 demonstrates that the effects of El Niño episodes (warm events) and the cool events, known as La Niña, occur across the entire globe.

There are also longer-term changes in ocean-atmosphere circulation – marked by shifts in the location and/or intensity of the semi-permanent high- and low-pressure cells. These changes can persist for several decades. For example, temperature and circulation patterns in the North Pacific appear to get "stuck" in one of two modes for long periods. Various indices provide measures of this tendency, but they all strongly depend on the intensity and position of the winter Aleutian low-pressure system. Figure 11 displays one such index: the Pacific Decadal Oscillation (PDO) Index. When the PDO is in its positive coastal warm phase, as it was for most of the period from 1977 through the mid-1990s, sea surface temperatures along the west coast of North America are unusually warm, the winter Aleutian low intensifies, and the Gulf of Alaska is unusually stormy.

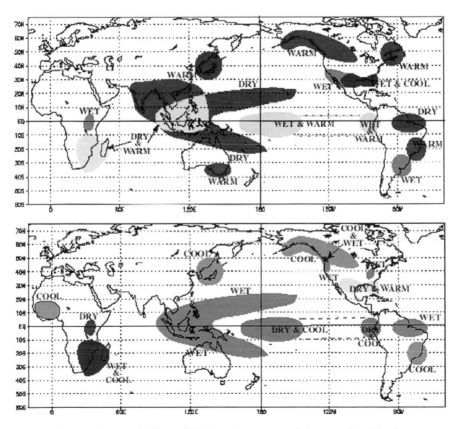

Figure 10. Expected seasonal effects of El Niño (warm episodes) across the globe during December–February (top) and expected seasonal effects of La Niña (cold episodes) during the same time period (bottom) (from Climate Diagnostics Center, NOAA).

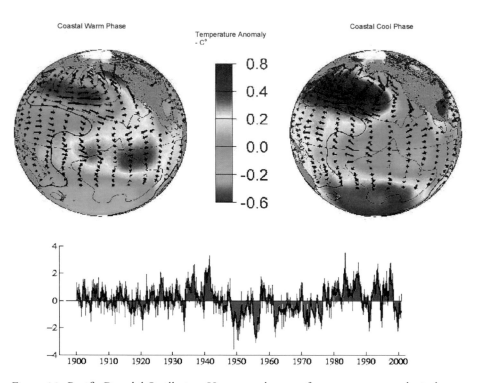

Figure 11. Pacific Decadal Oscillation. Upper panel: sea surface temperature and wind stress anomalies. Lower panel: Monthly values of PDO Index. Red is coastal warm phase; blue is coastal cool phase (courtesy of Dr. Nathan Mantua, JISAO, University of Washington and Stephen Hare, International Pacific Halibut Commission).

The slowly evolving state of the ocean, as measured by the PDO, interacts with the more rapid ENSO-related changes to influence storm tracks and, thus, the likelihood of unusually heavy or light seasonal precipitation. For example, a positive PDO appears to reinforce the effects of an El Niño, making wet winter conditions in the southwestern United States and dry conditions in the Pacific Northwest more likely than would be the case if the PDO were in the negative (coastal cool) phase.

A similar pattern of multi-year variability occurs in the Atlantic basin as well. The North Atlantic Oscillation (NAO) measures swings in the relative intensity of the winter low-pressure cell centered over Iceland, and the high-pressure cell centered over the Azores. The NAO is in a positive phase when that pressure difference is larger than normal. A positive NAO pattern drives strong, westerly winds over northern Europe, bringing warm stormy winter weather, while southern Europe, the Mediterranean and Western Asia experience unusually cool and dry conditions (Figure 12a). Also in the positive phase, northeastern Canada is more likely to experience unusually cold winter conditions. In the negative phase, the pressure differential is smaller than average and winter conditions are unusually cold over northern Europe and milder than normal over Greenland, northeastern Canada, and the Northwest Atlantic. There have been long periods during which the NAO has tended to be either unusually low or unusually high. In particular, it was generally low throughout the 1950s and 1960s, and then abruptly switched to a positive state for most of the period from 1970 to the present (Figure 12b).

ENSO, the PDO, and the NAO are all natural modes of climate variability, but any change in global climate is also likely to affect these processes. At the global scale, climate changes depend

Figure 12a. Schematic of the positive index phase of the North Atlantic Oscillation (NAO) during the Northern Hemisphere winter (courtesy of Dr. James Hurrell, CGD/NCAR).

on changes in the Earth's energy budget. In particular, increased concentrations of greenhouse gases, such as carbon dioxide, in the atmosphere are likely to cause warmer global average surface temperatures.

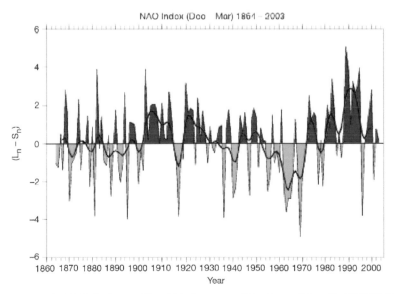

Figure 12b. NAO Index 1864–2003 (courtesy of Dr. James Hurrell, CGD/NCAR).

The Earth's climate has changed throughout geologic time – why did those changes occur?

There is strong evidence that the Earth has experienced long periods during which average global temperatures were much colder and much warmer than today. Changes in the Earth's climate system throughout geologic time can be linked to changes in the components of the climate system, including changes in the Earth itself, the composition of the atmosphere, and the seasonal distribution and total amount of incoming solar energy.

There have been enormous changes in the surface of the Earth – with continents moving, mountain ranges growing and eroding away, and the area covered by oceans and by ice growing or shrinking. The composition of the atmosphere has also changed as a result of biological and geophysical processes, including storage of carbon in the ocean and its subsequent release, volcanic eruptions, and the occasional sudden release of methane from sediments on the ocean floor. In addition, there have been changes in solar output, in the Earth's orbit, and Earth-Sun geometry. All of these changes affect climate at both the global and regional scale.

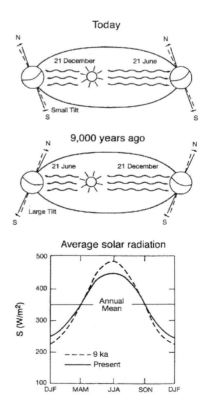

Consider, for example, the effects of slow changes in the Earth's orbit around the Sun. Over the course of approximately 100,000 years, the Earth's orbit around the Sun changes shape from a thin oval to a circle, and back again. At present, the shape of the Earth's orbit is almost a perfect circle. There is only a small difference in our distance from the Sun at the time when we are closest to it (the perihelion, currently in January), and when we are farthest away (the aphelion, currently in July). The fact that the Earth is now closest to the Sun during the northern hemisphere winter is just a coincidence, because the date of the perihelion slowly moves to come later in the year, following a 21,000-year cycle. In other words, 10,000 years from now, the perihelion will occur in the northern hemisphere summer, causing northern hemisphere seasonal contrasts to be somewhat more pronounced than at present (Figure 13).

Figure 13. Graphic illustration of the Earth's orbit and average solar radiation comparing present conditions with those 9 thousand years ago (9ka)(Trenberth et al. 2000).

Even such subtle differences can have profound impacts on regional climates. When the perihelion last occurred in the northern hemisphere summer, the Sahara was much wetter than it is now and was covered with savanna-like vegetation. As the seasonal distribution of solar radiation gradually changed to modern conditions, the Sahara dried out. Its transformation to the present-day desert accelerated dramatically about 5,500 years ago. The abruptness of the change suggests that the climate system crossed a threshold, triggering a series of biophysical feedbacks that amplified the trend toward regional drying (IGBP 2001).

Seasonal contrasts would also tend to be more extreme when the shape of the Earth's orbit is more elliptical than it is at present. In addition, the Earth wobbles slightly on its axis, so that the angle of the tilt varies over a 41,000-year cycle. Recall that the Earth's tilt causes seasons in the first place. So, the greater the angle of tilt, the stronger the seasonal contrasts. These astronomical Milankovich cycles appear to have played a significant role in the timing of ice ages and interglacial periods in the recent past, but they clearly cannot explain all of the Earth's climate history.

Changes in the seasonal distribution of incoming solar energy may have triggered the beginning and end of previous ice ages. However, the solar impacts were greatly amplified by positive feedbacks within the climate system, including changes in the reflection of sunlight back into space by ice-covered areas, changes in ocean circulation, and dramatic changes in atmospheric concentrations of greenhouse gases, especially carbon dioxide and methane. Over the past 400,000 years, the record of temperatures in the world's high-latitude regions followed a saw-toothed pattern. Global concentrations of carbon dioxide and methane followed a nearly identical pattern (Figure 14). There were four long but erratic periods of cooling, each followed by a dramatic warm-up. Scientists do not fully understand the reasons for this pattern, but changes in the ocean's thermohaline circulation (Figure 15) and changes in the release of carbon dioxide from the oceans, and the release of methane from wetlands, appear to have played important roles.

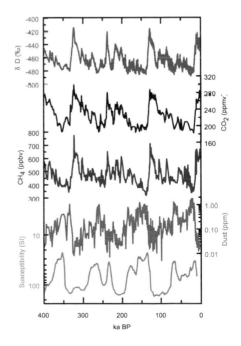

Figure 14. Four glacial cycles are recorded in Vostok ice cores. The graphic represents thousands of years before the present. The top three lines from the Vostok ice core record show Deuterium — a proxy for local temperature (blue); CO_2 (black); methane (red); and dust (purple). The green line is a measure of Chinese loess deposition. (after figure compiled by the PAGES program; K. Alverson et al., 2003)

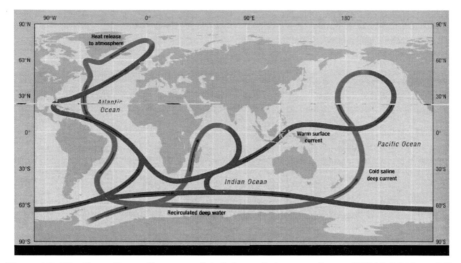

Figure 15. The Great Ocean Conveyor: global thermohaline circulation (WMO 2003, p. 52).

In Figure 14, one can see that rapid warming and increases in atmospheric carbon dioxide and methane occurred nearly simultaneously. This suggests a positive feedback loop, with initial warming causing the greenhouse gas concentrations to rise, and rising concentrations promoting further warming. Figure 14 also shows a correspondence between the temperature record and long periods of wet or dry conditions in Central and East Asia. Wind-borne dust deposits, both in Antarctica (Vostok) and on the Chinese Loess Plateau tended to peak during glacial periods, indicating expansion of Asian deserts.

Figure 15 depicts the approximate pattern of thermohaline circulation in the World's oceans – that is, the connection between the movement of cold, salty water in the oceans' depths and the movement of warm, less saline water at the surface (Broecker 1997). Warm, low-salinity water from the tropical Pacific and Indian Oceans flows around the tip of South Africa and ultimately joins the Gulf Stream to transport heat from the Caribbean to Western Europe. As the water moves northward, evaporative heat loss cools the water and leaves it saltier and more dense. The cold, salty water sinks in the North Atlantic and flows back toward Antarctica, thus pushing the conveyor along. One hypothesis is that the inflow of fresh water into the North Atlantic during warm periods can cause this conveyor to dramatically slow down or even collapse. Such a mechanism could explain the sudden reversals of warming that appear in the geologic record.

It is likely that increased high-latitude runoff and ice-melt caused by human-induced climate change will slow the thermohaline circulation. However, we do not know how much that would reduce projected temperature increases for Europe and the northern latitudes, because the mechanisms of human-induced climate change are different from the mechanisms of previous natural warming episodes (IPCC WG I 2001). This is an area of active research.

Why should I believe that emissions of carbon dioxide and other greenhouse gases will cause global climate change?

The major greenhouse gases, carbon dioxide, methane, nitrous oxide and water vapor, occur naturally in the atmosphere. Without them, the Earth would be too cold to support life as we know it. The basic science of the greenhouse effect is not controversial. Scientists understand the greenhouse effect and can easily reproduce it in the laboratory. There is no disagreement about the fact that these gases are transparent to incoming short-wave solar radiation, and that they tend to absorb outgoing long-wave radiation and re-emit part of that radiation back down to the Earth's surface. In effect, they act as a blanket to warm the surface of the Earth.

Concern about climate change arises from the fact that human activities are releasing large quantities of these substances – and other even more powerful manufactured greenhouse gases such as halocarbons – into the atmosphere (Table 2). Because carbon

Table 2

Selected chemically reactive greenhouse gases and their precursors: abundances, trends, budgets, lifetimes, and GWPs.

Chemical Species	Formula		Abundance		Trend- Annual % Change	Annual Emission	Atmo- spheric Lifetime	100-yr GWP[c]
		(units)	*2002*	*1750*	*1990s*	*late 1990s*	*(yr)*	
Carbon dioxide	CO_2	(ppm)	372	280	0.4 %	6.3 +/ - 0.4 PgC	~5 to 200	1
Methane	CH_4	(ppb)	1729[d]	700	0.4 %	600 Tg	12[a]	23
Nitrous oxide	N_2O	(ppb)	314	270	0.3 %	16.4 TgN	114[a]	296
Perfluoromethane	CF_4	(ppt)	80	40	1.3 %	~15 Gg	>50000	5700
Perfluoroethane	C_2F_6	(ppt)	3.0	0	2.7 %	~2 Gg	10000	11900
Sulphur hexafluoride	SF_6	(ppt)	4.2	0	5.7 %	~6 Gg	3200	22200
HFC-23	CHF_3	(ppt)	14	0	3.9 %	~7 Gg	260	12000
CFC-11[b]	$CFCl_3$	(ppt)	268	0	-0.5 %		45	4600
CFC-12[b]	CF_2Cl_2	(ppt)	533	0	0.8 %		100	10600

Sources: Data from IPCC WGI 2001; Blasing and Jones 2003.

[a] Species with chemical feedbacks that affect the duration of atmospheric response – values are perturbation lifetimes

[b] Regulated under Montreal Protocol

[c] Global Warming Potential (GWP) is an index describing the relative effectiveness of well-mixed greenhouse gases in absorbing outgoing infrared radiation. The index approximates the time-integrated warming effect of a unit mass of a given greenhouse gas relative to that of carbon dioxide.

[d] As measured at Cape Grim, Tasmania (Blasing and Jones 2003).

dioxide and many of the halocarbons have very long atmospheric lifetimes, the increased concentrations are likely to result in an enhanced greenhouse effect for centuries to come.

> *The basic science of the greenhouse effect is not controversial.*

We are also loading the atmosphere with other types of pollutants. Some of these tend to produce cooling by reflecting incoming sunlight. Dust from disturbed soil surfaces and other tiny particles from combustion, especially sulphate aerosols, act in this way. Unlike carbon dioxide and many other greenhouse gases, however, these aerosols only stay in the atmosphere a very short time. So, although they may temporarily mask the warming effects of the greenhouse gases, warming will eventually dominate. Figure 16 depicts the estimated relative impacts of greenhouse gases, aerosols, and other factors on global temperatures from pre-industrial times (circa 1750) to the present (circa 2000).

Over the past 400,000 years, atmospheric carbon dioxide concentrations varied from about 180 parts per million (ppmv) at the height of each glaciation to about 310 ppmv at the peak of each warming. Similarly, methane concentrations varied from approximately 350 to 800 parts per billion (ppbv). Since the beginning of the Industrial Revolution, burning of fossil fuels, deforestation, expanding agriculture, and other human activities have contributed to rapid increases in CO_2 and methane concentrations. In the mid-eighteenth century, the estimated atmospheric concentration of CO_2 stood at 280 ppmv. As of the year 2002, it had risen to

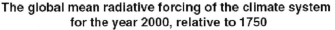

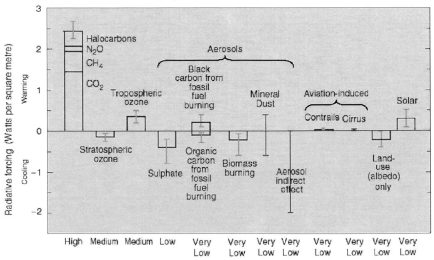

Figure 16. Many external factors force climate change. The error bars show ranges of uncertainty in radiative forcing (IPCC WGI 2001, p. 37).

approximately 372 ppmv. Similarly, methane concentrations increased from approximately 700 ppbv at the beginning of the Industrial Revolution to current levels between 1,729–1,843 ppbv, as measured at different locations. These modern levels are, thus, well above the range of natural variability in the recent geologic past. Future emissions are expected to further increase these concentrations (Figure 17).

What are the uncertainties regarding future climate changes?

There are really three big questions here: 1) How much warming is likely to result from a given scenario of human-caused increases in greenhouse gas concentrations? 2) What will that do to local and regional climates? 3) What will be the actual amounts of greenhouse gases added to the atmosphere in the future?

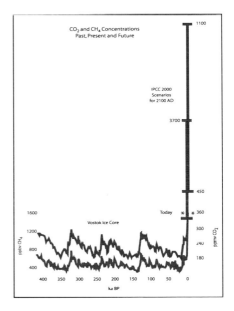

Figure 17. Carbon dioxide (CO_2) and methane (CH_4) concentrations: past, present, and future. (After figure compiled by the PAGES program: K. Alverson et al. 2003.)

The Third Assessment Report of the Intergovernmental Panel on Climate Change (IPCC) estimates that global average temperature will rise by between 1.4° to 5.8°C by the year 2100. This rather wide range of uncertainty results primarily from the fact that it is difficult to forecast future emissions, and also from the fact that the ultimate warming will depend on the size and direction of many feedback processes in the climate system that cannot be precisely estimated. Changes in atmospheric water vapor and cloud formation are two of the most important processes in this regard. The warming from increased CO_2 will be strongly amplified by associated increases in atmospheric water vapor, while changes in the extent of cloud cover and the characteristics of clouds may either enhance or diminish the initial warming. Accounting for the range of uncertainty in these feedbacks results in a range of possible changes in global average temperatures for any given change in CO_2, and in the other greenhouse gases.

Future greenhouse gas emissions are the real wild card because they depend on how fast the world economy grows, how fast world population increases, how quickly our energy technology evolves, and how much our land uses change. Most importantly, greenhouse gas emissions will depend on the policies we put in place to reduce the amount of climate change that will eventually occur.

Figure 18 presents a range of possible future paths for CO_2 emissions along with two different estimates of the resulting changes in the atmospheric concentrations of

CO_2.[1] One of the important things to note is that atmospheric CO_2 concentrations will be higher in the year 2100 than they are now, even in the scenarios in which emission rates eventually decline significantly relative to present rates. This indicates that some climate change will be inevitable. In fact, even if atmospheric CO_2 concentrations could be held fixed at today's levels, global mean temperature and sea levels would continue to rise for several centuries due to the thermal inertia of the oceans. This is commonly called "committed global warming," which Wigley (2005) estimates could range from 0.2 to more than 1°C. In addition, recall that CO_2 has a long atmospheric lifetime, and that emissions cannot be avoided completely – even under the most optimistic assumptions about future innovations in energy technology.

The other important thing to notice is that there are huge differences in projected CO_2 concentrations at the end of the century, depending on the development path followed by the world economy and future population growth. Also note that there is some uncertainty about the eventual CO_2 concentrations that

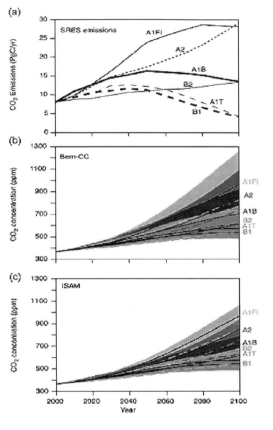

Figure 18. A possible range of carbon dioxide emissions and the resulting atmospheric changes. (IPCC WGI 2001, p. 222)

---

[1] The SRES emissions scenarios pictured here were developed as part of the IPCC 2001 assessment process. They represent a wide range of possible futures, as follows:

A1Fl = rapid economic growth, continued reliance on fossil fuels, converging world living standards, world population peaking in mid century and declining thereafter.

A1T = Same as above except with increasing reliance on new technologies using renewable energy rather than fossil fuels

A1B = Same as above except with a balance of fossil and non-fossil fuel sources

A2 = regionally divergent economic growth, continuing population growth, slower and more fragmented technological change

B2 = emphasis on local solutions to economic, social and environmental sustainability, intermediate technological change, economic growth and population growth

B1 = population as in A1, rapid change toward service and information economy, emphasis on clean, highly resource-efficient technologies.

would result from any given emission scenario – arising largely from our incomplete understanding of possible changes in the uptake and release of carbon by biological processes on both land surfaces and in the ocean.

Scientific understanding of the sensitivity of the climate system to projected changes in the concentrations of CO_2 and other trace gases is also imperfect.

> *Future greenhouse gas emissions are the real wild card... Most importantly, greenhouse gas emissions will depend on the policies we put in place to reduce the amount of climate change that will eventually occur.*

Different climate models will produce different projected temperature changes because they incorporate different estimates of the parameters that describe the behavior of the climate system. The range of temperature changes projected by the IPCC reflects the combined effects of all of these sources of uncertainty. Figure 19 compares the range

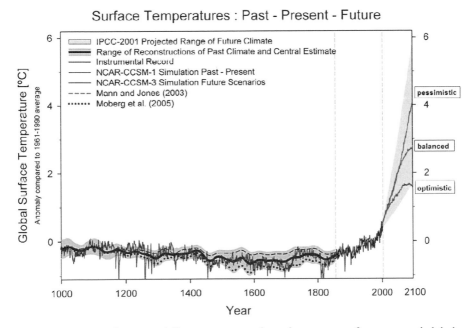

Figure 19. Many studies using different proxy records produce a range of reconstructed global temperature records for the past 1000 years. The range is roughly bounded by the low Moberg (2005) reconstruction and the high Mann and Jones (2003) reconstruction. Also shown is a climate model simulation of past climate based on geologic records of volcanic activity, solar variability, and estimated changes in greenhouse gas and aerosol concentrations. These are compared to the global instrumental record starting in the mid-nineteenth century and to projected temperature changes as of 2100 under three IPCC scenarios: A2 (pessimistic) A1B (balanced growth of fossil and non-fossil fuel use) and B1 (optimistic). (courtesy of Caspar Ammann, Climate and Global Dynamics Division, NCAR).

of IPCC temperature projections over the coming century with a range of estimated records of Northern Hemisphere average temperature changes over the past 1000 years. The shading represents the range of uncertainty in both the projections and the record of past variation.

Of course, global average temperature is a very crude metric of climate change. Nobody lives at the global average. What we really care about is what will happen to climate at particular places, and temperature is only one of several important variables. Water supplies, for example, will be affected by changes in temperature, precipitation (including changes in timing and intensity), insolation, humidity, and wind speed, among other factors. In addition, many human and natural systems are likely to be sensitive to changes in extremes (e.g., the frequency and severity of floods and droughts). Unfortunately, the details of climate changes at particular places and times in the future cannot be reliably projected at this time – even if we could reliably project the change in global average temperature. We will discuss this lack of certainty, and its implications for water utility planning below. Here, it is important to emphasize that while the details of local climate changes cannot be projected with high accuracy, we are beginning to accumulate some evidence on the likely characteristics of climate changes on gross regional scales. This body of evidence is sufficient to allow utilities to explore the implications of a range of potential local climate changes that are consistent with projections of global warming.

Is climate change really likely to happen on a time scale relevant to water utilities?

First, it is important to understand that climate change is already happening. Over the past century, global average surface temperature increased by approximately 0.6°C (See Figure 1). Warming is expected to accelerate during the current century. The modest warming to date has not been evenly distributed over the surface of the globe. In particular, arctic areas have warmed more rapidly than other areas. Climate model simulations also suggest that future warming will tend to be most pronounced in the higher northern latitudes (Figure 20). That picture might change if there is a significant slowing of the oceanic thermohaline circulation, because that would reduce the poleward transport of heat through the ocean.

As for impacts on water resources, warming over the past half-century appears to be associated with reduced spring snowpacks in some of the mountainous areas of the western United States. In California's Sacramento River

Over the past century, global average surface temperature increased by approximately 0.6°C. Warming is expected to accelerate during the current century.

34

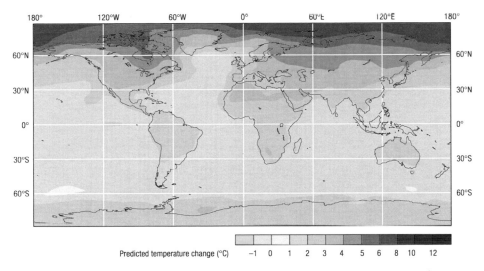

Figure 20. The multiple model ensemble map for the end of the 21st century projects that most warming will occur over the Arctic and land areas, when compared with the 1961–1990 Normals. (WMO 2003, p. 204)

Basin, for example, spring runoff has been peaking earlier, and there has been a century-long downward trend in late spring and early summer flow as a proportion of total annual flow (Dettinger and Cayan 1995). In addition, higher sea levels have caused saltwater intrusion problems in some areas. Over the past century, warmer temperatures contributed to rising sea levels through thermal expansion of the World's oceans and glacial melt (U.S. Geological Survey 2000). Measured increases vary at different locations as a result of local processes, such as subsidence. Miami-Dade County, for example, has experienced a twelve-inch increase in sea level since 1848.

Warming could have some benefits, couldn't it? Why is it usually portrayed as some sort of catastrophe?

Global climate change will certainly produce a mixture of both beneficial and harmful impacts. Beneficial impacts might include reduced winter heating demands and longer growing seasons in some areas, while harmful impacts will include the health and energy demand impacts of more frequent summer heat waves, and increased stress on natural ecosystems and poor countries that lack resilience to change. Here, we are particularly interested in the possible impacts of climate change on the water utility industry. As will be described below, large changes are possible in total available water supplies, in the seasonal distribution of surface flows, in water quality and in the frequency and severity of flood and drought events. However, the details of how these changes will unfold at any given location are likely to remain highly uncertain.

Water demands, particularly for irrigation, are likely to change as well. Effective adaptation to such changes may require careful evaluation of the implications of a wide range of possible future climate scenarios.

If we institute policies to slow down or stop the growth in emissions, how long would it take for climate to stop warming?

Climate change cannot be "turned off" immediately. The concentrations of greenhouse gases in the atmosphere and heat stored in the oceans, rather than current emissions, are what determine how warm the climate will be. To draw upon a water resource analogy, atmospheric concentrations of CO_2 can be thought of as a large reservoir, while current emissions are like a small stream entering the reservoir. The level of the reservoir rises or falls depending upon whether the natural draw-down processes (e.g. evaporation in the case of water) are larger or smaller than the current rate of inflow. There are natural processes by which CO_2 and other greenhouse gases are removed from the atmosphere (including uptake and storage of CO_2 in biota, soils, oceans and ocean sediments), but we are currently adding these gases much faster than they are being removed. In addition, these natural sinks may become saturated as CO_2 concentrations rise. On average, natural sinks currently remove over half of the carbon emitted by fossil fuel use each year, but these processes almost certainly will become less effective in a warmer world. For example, the solubility of CO_2 in seawater declines as the water warms (IGBP 2001), which would reduce the effectiveness of the ocean sink.

It is also important to understand that the thermal inertia of the oceans results in a time lag between changes in atmospheric greenhouse gas concentrations and changes in global average temperatures. Wigley (2005) estimates that climate would continue to warm and sea levels would continue to rise for several centuries, even if greenhouse gas concentrations were immediately stabilized. In other words, stopping climate change would require stabilizing atmospheric concentrations of greenhouse gases,[2] and then waiting for a long time while the climate system gradually equilibrated. The reduction in emissions that would be required to do that depends on the levels at which CO_2 and other greenhouse gas concentrations are to be stabilized, and the target date for that stabilization.

---

[2]Note that it is actually the radiative forcing of the entire suite of greenhouse gases that would have to be stabilized, so there may be several ways to achieve a mix of emission reductions to meet any specific stabilization target.

Hydrologic Implications for Water Utilities

Global climate change will likely alter the hydrologic cycle in ways that may cause substantial impacts on water resource availability and changes in water quality. For example, the amount, intensity, and temporal distribution of precipitation are likely to change. Warmer temperatures will affect the proportion of winter precipitation falling as rain or snow, how much is stored as snow and ice, and when it melts. Long-term climatic trends could trigger vegetation changes that would alter a region's water balance. In forested areas, the combination of warmer temperatures and drying soils caused by earlier snowmelt or longer drought periods could cause wildfires to become more frequent and extensive. Where that occurs, land cover and watershed runoff characteristics may change quickly and dramatically as wildfires reduce forest cover, thereby altering the runoff

> *The amount, intensity, and temporal distribution of precipitation are likely to change.*

response. Less dramatic but equally important changes in runoff could arise from the fact that the amount of water transpired by plants will change with changes in soil moisture availability, and plant responses to elevated CO_2 concentrations. In addition, changes in the quantity of water percolating to groundwater storage will result in changes in aquifer levels, in base flows entering surface streams, and in seepage losses from surface water bodies to the groundwater system. There is a rich scientific literature describing the potential influence of climate change on both individual water cycle components and the overall hydrologic cycle. This section of the Primer provides a brief summary of potential impacts on the most important elements.

Precipitation amount

A change that appears most likely is that global average precipitation will increase as global temperatures rise. Evaporation potential will increase with warming because a warmer atmosphere can hold more moisture. This capacity is governed by the exponential Clausius-Claperyon equation, which states that for a one-degree Celsius increase in air temperature, the water-holding capacity of the atmosphere is increased by about seven percent.

A simple-minded explanation for the resulting intensification of the hydrologic cycle is that "what goes up, must come down." Of course, it really is not that simple, but the overall scientific consensus is that globally the Earth will be warmer with higher globally averaged precipitation. Exactly how much global average precipitation will increase is less certain. On average, current climate models suggest an increase of about

1–2 percent per degree Celsius due to warming forced by CO_2 (Allen and Ingram 2002). An increase in global average precipitation does not mean that it will get wetter everywhere and in all seasons. In fact, all climate model simulations show complex patterns of precipitation change, with some regions receiving less and others more precipitation than they do now. The local balance between changes in precipitation and changes in actual evaporation will determine the net change in river flows and groundwater recharge.

In general, the models agree in projecting precipitation increases over high-latitude land areas, much smaller and less certain increases over the equatorial regions, and decreases over some subtropical areas. Elsewhere, precipitation changes are more variable across models (Carter et al. 1999; IPCC WGI 2001). Wigley (2004) has developed a statistical summary of the spatial distribution of precipitation change seen in scenarios generated by various climate models. Figure 21 displays these results in the form of normalized signal-to-noise ratios, where noise represents scatter among model projections. In other words, at the red end of the spectrum, the models tend to agree on increased precipitation. At the opposite end, where the map is shaded in blue tones, they tend to agree on reduced precipitation. However, in the middle of the color spectrum (corresponding to the green and yellow-green tones), the various projections give differing results regarding whether annual precipitation will increase or decrease. This suggests that mid-latitude areas such as the continental U.S. and much of Europe and Asia face an especially uncertain future regarding changes in average annual precipitation.

The major difficulty is that although different model simulations are fairly consistent in regional temperature changes, they often display very different

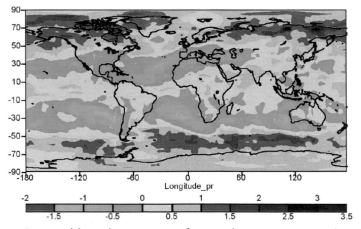

Figure 21. Inter-model signal-to-noise ratios for annual-mean precipitation (mean precipitation change per 1°C global-mean warming, averaged over 17 AOGCMs, divided by the inter-model standard deviation). This is a measure of both the sign and strength of the expected precipitation change and the level of agreement between models. Values between –1 and +1 indicate considerable uncertainty in the expected change. Source: Wigley 2004. The Benefits of Climate Change Policies: Analytical and Framework Issues, *copyright OECD, 2004.*

regional precipitation patterns. To understand why this occurs and what it implies for the usefulness of climate model projections, it is helpful to begin with an explanation of what climate models are and how they are used to simulate present and future climates.

Mid-latitude areas face an especially uncertain future regarding changes in average annual precipitation.

Coupled Atmosphere-Ocean General Circulation Models (AOGCMs) are currently the primary tool used to analyze the potential impacts of increased greenhouse gases, aerosols and other factors on global climate. To be useful for the analysis of climate change, the atmospheric model must be coupled to models of other components of the climate system, such as the oceans, the sea ice, and the land surface. The major climate models include tens of vertical layers in the atmosphere and the oceans, dynamic sea-ice sub-models, and effects of changes in vegetation and other land surface characteristics (Gates et al. 1996; Washington 1996). The atmospheric part of a climate model is a mathematical representation of the behavior of the atmosphere based upon the fundamental, non-linear equations of classical physics. A three-dimensional horizontal and vertical grid structure (as depicted in Figure 22) is used to track the movement of air parcels and the exchange of energy and moisture between parcels.

Despite tremendous technological advances in computing capability, it is still very time-consuming and costly to use these models to simulate future climates. One of the most important choices for achieving model results in a reasonable amount of time is to increase the model's horizontal resolution. This limitation means that it is prohibitively costly to run full coupled-climate models at a spatial resolution that would accurately depict the effects of mountains and other complex surface features on regional climates.

The problem with such a coarse horizontal resolution is that important processes that occur at finer scales are not

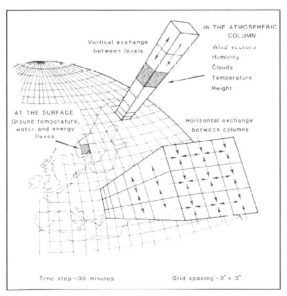

Figure 22. The structure of an atmospheric GCM.
Source: Henderson-Sellers and McGuffie 1987. A Climate Modelling Primer. *Copyright John Wiley & Sons Limited. Reproduced with permission.*

well resolved. Topography, for example, is very important in determining the location of precipitation. As moist air rises over mountains or hills, the moisture condenses, producing clouds and if conditions are right, precipitation occurs. Although there has been marked improvement over the last three decades, the coarse horizontal resolution of typical climate models still tends to smooth out important landscape features that affect atmospheric processes. At the resolution of most AOGCMs, the models see the mountains of the western United States as a large set of ridges and do not resolve finer-scale features that influence regional climate. Clearly, that spatial resolution is too coarse to reproduce the effects of topography on the region's precipitation and runoff patterns (Grotch and MacCracken 1991; Giorgi and Mearns 1991; Pan et al. 2004). For example, coarse-resolution models would see the Great Basin area as being located on an upslope. They would therefore predict it to be wet, when it is actually a desert. The global-scale models cannot adequately capture the actual rain-shadowing effect of the Sierra Nevada Mountain Range. In short, raw AOGCM output will put the precipitation in the wrong places and perhaps at the wrong time.

Recognition of limits imposed by the relatively coarse horizontal scales of AOGCMs has led to the application of "downscaling" as a means of trying to understand how local-scale processes, of greater interest to water utilities, might respond to larger-scale weather and climate changes (Wilby, et al., 2004). Downscaling includes statistical methods and the use of regional climate models run at a relatively high resolution over a limited area with boundary conditions (and sometimes interior domain information as well) prescribed from the lower resolution AOGCM. Like global climate models, regional climate models will vary in their precipitation projections depending on the downscaling method, the model specifications, and the AOGCM scenario that is downscaled. While it is possible for a downscaled model to resolve some limitations of general circulation models for a specific region, they are still limited in their capabilities to give reliable projections for future precipitation. Downscaling can produce more sub-regional detail but not necessarily more information. The section entitled "Climate Change Information in Utility Planning" provides a further discussion of the usefulness of downscaled models in generating plausible scenarios for use by water utilities.

Precipitation frequency and intensity

In addition to changes in global average precipitation, some have argued there could be more pronounced changes in the characteristics of regional and local precipitation due to global warming. For example, Trenberth et al. (2003) hypothesized that, on average, precipitation will tend to be less frequent, but more intense when it does occur, implying greater incidence of extreme floods and droughts, with resulting consequences for water storage. Their arguments are based on the premise that local and regional rainfall rates greatly exceed evaporation rates and thus depend on the convergence of regional to continental scale moisture sources. They surmise that rainfall intensity should increase at about the same rate as the increase in atmospheric moisture, namely 7 percent per degree Celsius with warming. This means that the changes in rain rates, when it rains, are at odds with the 1–2 percent per degree Celsius model estimates for total rainfall amounts as discussed previously. The implication is that there must be a decrease in light and moderate rains, and/or a decrease in the frequency of rain events, as found by Hennessey et al. (1997). Thus, the prospect may be for fewer but more intense rainfall – or snowfall – events.

Evaporation and transpiration

Evaporation from the land surface includes evaporation from open water, soil, shallow groundwater, and water stored on vegetation, along with transpiration through plants. The combined effect, commonly referred to as evapotranspiration, has a substantial influence on basin water budgets, runoff, and groundwater recharge. There is an enormous hydrologic literature regarding the nature, response, and controls of evapotranspiration under current and future climate conditions, but the interplay between atmospheric energy, moisture, and turbulence, and plant water use efficiency under different water, energy, nutrient, and CO_2 levels is complex and not yet fully understood.

A consistent prediction of climate models is that global warming will increase total evaporation. Increases in surface temperature and higher wind speeds promote potential evaporation, while the greatest change will likely result from an increase in the water-holding capacity of the atmosphere. While potential evaporation will almost certainly increase with temperature, its impact on precipitation in specific regions remains uncertain. There are many balances and counter-balances in the atmosphere that aren't fully understood. For example, atmospheric moisture originating from actual evaporation over oceans may help offset, and possibly even lessen, potential evaporative pressures over land. Likewise, there are regional controls on evaporation. In humid regions where water is not limiting and actual and potential evaporation are nearly equal, evaporation is constrained by the water-holding capacity of air above the surface, so an increase in this capacity due to warming may have a large evaporative effect. In dry regions, other factors such as surface water availability, surface temperature and wind are more important determinants of actual evaporation. A reduction in summer soil water, for example, could lead to a reduction in the rate of actual evaporative demands from a catchment despite an increase in potential evaporation. Arnell (1996) estimated for a sample of UK catchments that the rate of actual evaporation would increase by a smaller percentage than the atmospheric demand for evaporation, with the greatest difference between actual and potential evaporation occurring in the "driest" catchments, where water limitations are greatest (IPCC WGII 2001).

In their 1993 study on the Colorado River Basin, Nash and Gleick demonstrate the importance of evapotranspiration in determining water availability. The study modeled the impact of several climate change scenarios on runoff. Hypothetical precipitation changes ranged from a 20-percent increase to a 20-percent decrease, and the study considered temperature increases of 2°C and 4°C. Their study showed that changes in precipitation would cause proportional changes in runoff, if all else remained constant. Therefore, an increase or decrease in precipitation of 20 percent would result in runoff changing by approximately 20 percent. However, the impact of temperature on runoff was also substantial, due to evapotranspiration. The study found that, with no change in precipitation, a 2°C increase in temperature would reduce mean annual runoff by 4 to 12 percent. The change in runoff for a 4°C increase would be between 9 and 21 percent. Therefore, if temperature increased by 4°C, precipitation would need to increase by nearly 20 percent to maintain runoff at historical levels.

Changes in average annual runoff

Runoff changes will depend on changes in temperatures and precipitation, among other variables. A study by Arnell (2003) used several climate models to simulate future climate under differing emissions scenarios. The study linked these climate simulations to a large-scale hydrological model to examine changes in annual average surface runoff by 2050 (Figure 23). The striking thing about this figure is the fact that all simulations yield a global average increase in precipitation (not shown), but likewise exhibit

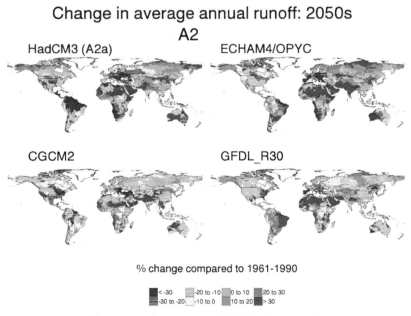

Figure 23. Percentage changes in average annual runoff projected by four climate models under IPCC Scenario A2 (Source: Courtesy of Nigel Arnell).

substantial areas where there are large decreases in runoff. Thus, the global message of increased precipitation clearly does not readily translate into regional increases in water availability. In addition, the fact that these different simulations produce quite different regional impacts demonstrates the uncertainty related to climate projections. In North America, for example, some models project much larger areas with reduced runoff than do other models.

What about natural variability?

It is very important to understand that natural variability will not go away. Any projected change in average annual runoff will occur "on top" of ongoing natural variability. In many cases, natural variability can be quite large compared to the changes projected from global warming. Furthermore, relatively short instrumental records may not provide an adequate picture of the full range of natural climatic variability. The work of several researchers who have developed proxy records for precipitation and streamflow based on tree rings and geological evidence provides a longer-term view.

Figure 24 provides examples of such proxy records for the reconstructed streamflow of the Colorado River at Lee's Ferry and for the Four Rivers Index in northern California (Sacramento, American, Yuba, and Feather). These 20-year moving averages indicate that both regions have experienced extended periods of drought as well as periods of sustained above-average flow. While there is a very weak positive correlation

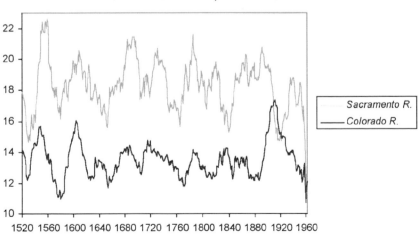

Figure 24. Time series plots of 20-year running means of reconstructed flows for the Colorado River at Lee's Ferry (lower line) and for the Four Rivers Index, northern California (upper line). Data Sources: Stockton and Jacoby 1976, 2004; Meko et al. 2001a,b – available at: http://www.ncdc.noaa.gov/paleo/recons.html.

between annual flows in the two regions, there is no consistent pattern of association for the longer-term fluctuations between wet and dry conditions. For example, northern California experienced an extended dry period from 1918–37, during which time the Four Rivers Index dropped to 13.55 million acre-feet (Maf) from its long-term mean of 17.4 Maf. At the same time, conditions in the Upper Colorado River were much wetter than the long term mean of 13.5 Maf. On the other hand, the most severe extended drought in the Upper Colorado River Basin occurred during the period 1579–98, when average annual flow was only 10.95 Maf. That same period was among the driest in the northern California tree-ring record (Meko et al. 2001 a,b). Paleo-environmental records also indicate that there were some "mega-droughts" in the pre-instrumental period that were far more severe than any experienced within recent history (Woodhouse and Overpeck 1998).

> *Water supplies can change dramatically, and for extended periods, even without anthropogenic climate change.*

These records suggest that water supplies can change dramatically, and for extended periods, even without anthropogenic climate change. Where they are available, such reconstructions of past variability could be useful for examining the vulnerability of a water system to conditions outside of the range of recent experience.

Temperature, snowpack and runoff

There is a high level of confidence in projections of warmer temperatures over most land surfaces. Unlike their projections of precipitation change, climate models are fairly consistent in predictions of regional surface temperature. Because temperature is central in determining the accumulation and melting of snow and ice, these scenarios are especially relevant to regions where snowpack or glacial runoff dominate the hydrology. In a warmer climate, it is very likely that a greater portion of winter precipitation will fall as rain rather than snow, especially in areas where winter temperatures are now only slightly below freezing. An increase in rain events would increase winter runoff but result in smaller total snowpack accumulations. Temperature also determines the timing of melt-off, and a warmer climate will likely result in an earlier melt season. Many regions are likely to see an increase in winter or spring flows and reduced summer flows. In fact, there is evidence that this is already occurring. Studies by Cayan et al. (2001) and Stewart et al. (2004) document the fact that the peak in spring runoff has been arriving earlier in the last few decades (Figure 25).

Warmer temperatures could increase the number of rain-on-snow events in some river basins, increasing the risk of winter and spring floods (Lettenmaier and Sheer 1991; Hughes et al. 1993). In currently glaciated basins, declining glacier reservoir capacity may eventually lead to an earlier peak of seasonal runoff and reduced late-

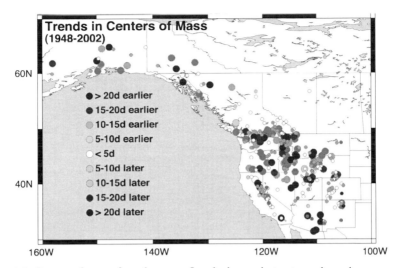

Figure 25. Centers of mass of yearly streamflow hydrographs in rivers throughout western North America, based on US Geological Survey streamflow gaging stations in the United States and and an equivalent Canadian streamflow network. Large circles indicate sites with trends that differ sigificantly from zero at a 90% confidence level; small circles are not confidently identified. (Courtesy of Michael Dettinger, based on Stewart et al. 2005.)

summer streamflows. In some cases, increased melting of glacial ice can sustain summer streamflows in the near term but will deplete this source in the long run (Pelto 1993).

The loss of snow mass from sublimation (sublimation is the change of ice to water vapor, bypassing the liquid phase) is a critical part of basin-scale water budgets in snow-dominated regions, but is an understudied topic. In exposed landcover regions such as prairie and tundra, Pomeroy and Gray (1995) estimated sublimation loss of blowing snow to be 15-41 percent of annual snowfall. Sublimation attributed to radiative energy tends to be greater in areas with less cloud cover (e.g., the Southwestern US), as sublimation is enhanced under direct sunlight, since photons of solar energy add the energy necessary for solid ice molecules to escape. On the eastern slopes of the Rocky Mountains, the warm and dry Chinook or "snow-eater" winds will quickly sublimate a snowpack, leading to unexpected reductions in basin water budgets. It is unclear how climate change could affect sublimation dynamics, since all three forces that contribute to sublimation (solar forcing, wind, and blowing snow) could change under anthropogenic warming.

Coastal zones

The IPCC Working Group II (2001) Third Assessment Report identifies sea level rise as one of the most important coastal impacts of global warming, and identifies several key impacts. A number of these are particularly relevant for water utilities located in coastal areas, including: 1) lowland inundation and wetland displacement; 2) altered tidal range in rivers and bays; 3) changes in sedimentation patterns; 4) severe storm-surge flooding; 5) saltwater intrusion into estuaries and freshwater aquifers; and 6) increased wind and rainfall damage in regions prone to tropical cyclones.

These impacts are particularly likely to affect water utility infrastructure. For example, there could be impacts on water intakes located in transition areas between freshwater and saltwater interfaces of both surface and sub-surface systems. Sedimentation patterns in estuaries and deltas depend strongly on tidal patterns, storm surges and flow conditions, whose changes could affect utility supplies.

Saltwater intrusion into freshwater aquifers is already a problem in many coastal communities, primarily due to overdrafting of those groundwater supplies. Because of the higher density of saltwater, a rise in sea level could result in a disproportionate loss of freshwater aquifers in coastal zones due to the intrusion of the saltwater wedge.

Case Study: Climate Change and Coastal Aquifers – Miami, FL and The Netherlands

Coastal regions such as Florida and the Netherlands are susceptible to various impacts of climate change, although sea level rise is arguably the greatest concern for these low coastal regions. Several studies have examined recent historical, and projected future, sea level rise resulting from melting glaciers and thermal expansion of the oceans as they become warmer (Levitus et al. 2000; USGS 2000; Meehl et al. 2005). This issue is familiar to areas such as Miami-Dade County, which has measured a twelve-inch rise in sea level since 1848. If the historical average rate continues, sea level will increase another three inches by 2025. Global warming will accelerate the rate of sea level change. Miami-Dade County has responded to this possibility by anticipating an increase of five inches by 2025, and a comparable rise in sea level over the remainder of the century. Likewise, the Netherlands is considering increases in sea level that, if realized, would significantly affect the nation's freshwater supply.

Sea level rise can have several impacts on coastal utilities. The most visible effect is damage to freshwater infrastructure caused by flooding, but also significant is saltwater intrusion into coastal aquifers, such as the Biscayne Aquifer that supplies Miami-Dade's 2.2 million inhabitants. To assess these issues, in 2002 the South Florida Regional Planning Council began mapping effects of sea level rise for coastal counties. These maps reveal that an increase in sea level of only five inches would be enough to inundate some of Miami's freshwater facilities. Key among the decisions Miami's water department must make in the near future will be whether to invest in protecting vulnerable facilities through flood control and drainage infrastructure or abandon them. In October 2003, the County established a task force to identify technically sound and economically viable responses in infrastructure planning to cope with sea level rise and other regional climate change impacts in the 21st century.

The second impact of sea level rise, saltwater intrusion, has been apparent long before climate change became a recognized threat. At the beginning of the twentieth century, saltwater did not intrude into the Biscayne Aquifer beyond the coast, but extensive construction of drainage canals provided an inlet for saltwater into the aquifer. Construction of control structures to hold back saltwater began in the 1940s, and more recently saltwater intrusion has been partially stabilized. However, groundwater is still contaminated several miles inland of the coast. This intrusion has driven the location of well fields and treatment facilities inland. Many of Miami's wells are located far enough inland that a rise of several inches in sea level will result in the loss of several inches at the base of the aquifer, which is small considering that the depth of the aquifer ranges between 100 to 150 feet. However, constructing wells inland has come at the cost of competing with the Everglades for fresh water. Environmental regulations protect the Everglades, especially since recent efforts have begun to restore the region's ecology. Therefore, the groundwater supply for Miami's water utilities will be constrained by both encroaching saltwater from the coast and limits on the utility's ability to continue moving its well fields inland due to the environmental needs of the Everglades. Additional constraints are the inability to maintain high enough water levels at the salinity control structures during droughts, and the need to open the salinity control structures and release water to prevent inland floods during high rainfall periods.

Saltwater intrusion will be confined within several miles of the coast in Florida, but regions that are already below sea level, such as parts of the Netherlands, will face more drastic impacts from contaminated groundwater. In the Netherlands, enhanced saltwater intrusion will result from both sea level rise and a shortage of fresh surface water to maintain water levels in polder (reclaimed, low-lying) areas during extreme dry summers. This may result in saltwater contamination of fresh groundwater aquifers. Since sixty-five percent of the Dutch drinking water supply comes from groundwater, it is obvious that climate change and saltwater intrusion may affect drinking water supply over the course of this century. Twenty-five percent of the Netherlands' 200 drinking-water facilities are situated just above or below sea level. At many of these sites, groundwater is vulnerable to saltwater contamination. Fifteen percent of the 200 facilities are estimated to be threatened due to saltwater intrusion and up-coning of saltwater from deeper (fossil) marine aquifers, and fifteen of the country's 200 freshwater production sites have already closed because of saltwater contamination. Such a possible drastic reduction in available groundwater in the future will make the Netherlands much more dependent on surface water, which currently accounts for only thirty-five percent of freshwater production.

The Netherlands does have an ample supply of surface water, but it is often polluted. The Netherlands' surface supply originates from Belgium, Germany and France via the rivers Rhine and Meuse, so contaminants from other countries' industry, agriculture, and urban regions inevitably end up in the Netherlands' water. Because the source of much of the water pollution lies outside of the Netherlands, it is rarely feasible to prevent contamination through regulations, although tackling such pollution is a goal of the EU Water Framework Directive (adopted in 2000). Therefore, the Dutch water utilities will need to incur the expense of sanitizing surface water. If a fifteen percent decrease in groundwater supply (including riverbank filtrate) occurs during this century, Dutch utilities will be required to filter over 100 billion liters of additional surface water per year to meet current demands. Purifying such a large quantity will be an expensive task and one that will take some time to implement, particularly if warmer water temperature and variable river flows further impair the already poor quality of the country's surface water. For this reason, the Dutch utilities carry out research on the production of drinking water from brackish groundwater using reverse osmosis membranes and on prevention of saltwater contamination of fresh aquifers by experimental well field design.

Water quality

Several aspects of climate change may lead to impacts on water quality. There is a consensus that the broad-scale hydrological cycle will intensify as the climate warms, with water quality adversely affected by the impacts of warmer temperatures, increased frequency of low-flow conditions, and possible increases in the intensity of episodic high-precipitation events. These two extremes of the hydrologic cycle, flooding and drought, pose potential threats to water quality.

At one extreme, heavy precipitation events may result in increased sediment and non-point source pollutant loadings to watercourses. This may make water treatment more difficult. Floods, in particular, increase the risk of water source contamination from sewage overflows, and runoff from agricultural land and urban areas. The location of water infrastructure, including both intakes and pipe distribution networks, could be increasingly vulnerable to precipitation extremes. Physical damage to dams and water operations and treatment facilities is a possible consequence of severe floods. Regions with combined sewage and storm runoff systems could have more frequent sanitary control problems due to flooding.

At the other extreme, where streamflows and lake levels decline, water quality deterioration is likely as nutrients and contaminants become more concentrated in reduced volumes with longer water residence times. Warmer water temperatures may have further direct impacts on water quality, such as reducing dissolved oxygen concentrations. Cold-water species, such as most salmon and trout, are particularly susceptible to warm water temperatures, and increasingly frequent warm water conditions could bring new challenges to the way managed river systems are controlled. In addition, evaporative water losses could increase the salinity of surface waters, especially in lakes and reservoirs with long residence times. These stresses on water quality will increase if climate change leads to longer dry spells. Contaminants tend to accumulate on land surfaces during prolonged droughts. Pulses of contaminated runoff can occur when precipitation returns. Water quality impacts are, therefore, likely to be rather complex and will vary with the physical, geographical and biological details of each water supply.

Case Study: Climate Variability and Water Quality – New York, NY

New York City depends on an unfiltered surface supply to provide 9 million consumers with approximately 1.5 billon gallons of water per day. The supply consists of nineteen cascading reservoirs and three controlled lakes located in the Catskill-Delaware and Croton catchments encompassing 5100 square kilometers (i.e., 1972 square miles) Figure 26. The only forms of treatment are chlorination, fluoridation, and corrosion control. In order to meet filtration avoidance requirements, coliform bacteria, turbidity, temperature, pH, oxygen content, and other water quality indicators must be monitored at least daily. Climate change has the potential to affect such water quality characteristics, and New York City's Department of Environmental Protection has identified the deterioration of water quality as a potential vulnerability to climate change.

The effect of climate change on turbidity is one of the most significant concerns for New York's water supply, as federal regulations limit it to 5 NTU (nephelometric turbidity units) at the intakes. Currently, typical values range between 0.5 and

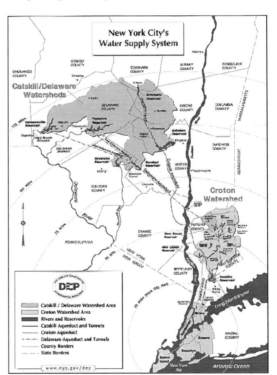

Figure 26. New York City's Water Supply System. Source: Courtesy of New York City Department of Environmental Protection.

1.5 NTU. Extreme weather events, such as hurricanes and nor'easters, can dramatically increase the turbidity levels at upstream reservoirs. Heavy rains that erode streambeds or sedimentary deposits from the last ice age transport glacial clays into the water supply, resulting in high turbidity. The flooding caused by Hurricane Floyd (September 18, 1999; see Figure 27), Hurricane Ivan (September 17, 2004), and heavy spring rain on April 2, 2005 resulted in turbidity values in the Ashokan Reservoir between 300 and 500 NTU. If heavy rains occur when water levels are low, as they are during droughts,

Figure 27. Schoharie Reservoir after Hurricane Floyd. Source: Courtesy of New York City Department of Environmental Protection.

exposed shorelines are vulnerable to severe erosion that can also result in high turbidity. When peak turbidity levels occur, NYCDEP may treat up to approximately 600 MGD in the Catskill Aqueduct with alum and sodium hydroxide to precipitate the clay particles and remove them from water traveling to the intakes. Water quality monitoring is greatly intensified and reported to the state regulatory agencies on a daily basis. Both the physical endurance of the staff and structural fitness of the system are tested at these times. An increased frequency of such events in the future would most likely necessitate increased staffing and intensified maintenance of equipment.

In order to reduce chemical treatment needs caused by extreme conditions, other turbidity reduction programs are in progress. NYCDEP is conducting a study of structural (e.g., intake design, turbidity curtains, etc.) and non-structural (e.g., operational) alternatives to control turbidity leaving Schoharie Reservoir. Other turbidity sources, such as suburban developments, are controlled by stormwater Best Management Practices (e.g., detention basins) to reduce turbidity in storm runoff at key locations near intakes. In addition, the City is currently implementing a Stream Management Program in the Catskill Mountains to reduce streambed and streambank erosion during stream baseflow using a geomorphic approach developed by Rosgen (1996). The Stream Management Program is most likely to be effective in controlling turbidity at low flows.

In addition to turbidity events, temperature change will have some important impacts on the operation of the NYC water supply. The reservoir system is a network of interconnected waterbodies linked by natural streams and aqueducts. Warm water results in more rapid settling of turbidity with the consequence that the best water (i.e., lowest turbidity) to send towards distribution is generally somewhat warmer than the high turbidity water that should be detained upstream until settling occurs. In an effort to use the best quality water, it may become increasingly difficult

Figure 28. Algae are a major source of disinfection byproducts (DBPs) that are regulated in drinking water. Source: Courtesy of New York City Department of Environmental Protection.

to balance the need to maintain low temperatures in the releases to streams (which are regulated to maintain cold-water fisheries habitats) with the quantity and quality requirements for the drinking water supply.

Another possible consequence of increases in rainfall may be increases in nutrient loadings to reservoirs and subsequent eutrophication. High phosphorous levels occur in reservoirs close to farmland. Elevated phosphorus concentrations can cause extensive blue-green algal blooms (see Figure 28) that contribute high levels of organic compounds to the drinking water. These organic compounds are pre-cursors of disinfection by-products (DBPs). On the distribution side of this problem, the Department of Environmental Protection (DEP) has conducted research to determine the optimum chlorine dose under various conditions to minimize DBP formation, while meeting the contact time requirements for disinfection.

Increased water temperature affects not only concentrations of suspended sediments, but also biological agents, such as coliform bacteria and water-borne pathogens. One possible explanation is the influence of temperature on sinking rates. DEP has observed that particles and pathogen cysts appear to diminish in reservoirs as water temperature increases, but more extensive analysis is needed to define this relationship.

Temperature can also affect survival and distribution of many microorganisms, their hosts, and their predators. As an example, it may change the behavioral patterns of migrating waterfowl, such as Canada geese, that can have a major influence on fecal coliform levels in reservoirs. There is a strong and well documented positive correlation between the number of waterbirds roosting on Kensico Reservoir (that reach a peak during seasonal migration periods) and the percentage of fecal coliform samples above the Surface Water Treatment Rule (SWTR) limit. DEP currently conducts a Waterfowl Management Program to keep geese, ducks and gulls away from intake areas. This maintains fecal coliform bacterial concentrations at low levels and within regulatory limits.

Although *Cryptosporidium* and *Giardia* have not historically been a significant problem for NYC, the utility is studying the sources and behavior of these water-borne pathogens. Knowledge of pathogen sources and behavior will allow DEP to develop effective management programs to mitigate the indirect effects that temperature change or intensification of the hydrological cycle may have on water quality. For example, a DEP study used genotyping of *Cryptosporidium* oocysts from a stream draining a residential area to demonstrate that nearly 90% of the cysts were of non-human origin. Microbiological "fingerprinting" studies such as this can help identify pathogen sources, indicate their importance for human consumers, and guide effective control measures. These studies are excellent examples of how utilities can reduce future risks by assessing potential vulnerabilities to climate-related impacts and acting to reduce these vulnerabilities.

In 2002, NYCDEP became one of 17 partner organizations in the CLIME Project (Climate Impacts on Lakes in Europe) sponsored by the European Commission. This three-year project is investigating the impacts of future climate change on water supply. An overview of the project is given at the website www.water.hut.fi/clime. The CLIME project was designed to analyze the impacts of climate change on freshwater resources. If present trends continue, limnologists believe that weather will have a major effect on the dynamics of lakes and reservoirs, including climate-related problems such as increased productivity, increased color, and increased frequency and severity of algal blooms. The benefits to New York City include an exchange of scientific expertise that broadens NYCDEP's current capabilities, particularly in the realm of prediction of future regional weather that will determine hydrologic conditions. It also allows first-hand involvement in the development of CLIME models and a decision support system that will lead to effective watershed management and planning for the future.

Most recently, in 2004, the NYCDEP instituted an agency-wide Climate Change Task Force (CCTF). The mission of the Task Force is to understand how climate change may affect the water supply and its infrastructure, and to provide a basis for long-term planning. The potential effects include sea level rise, temperature rise, an increase in extreme events, and changing precipitation patterns, all of which will have significant impacts on the City's existing water supply and wastewater treatment systems. Future infrastructure will also have to take these changes into account. An interesting approach taken by the CCTF has been to use extreme events (hurricanes and droughts) to begin to quantify future needs. Infrastructure changes may take decades to implement and therefore, advance planning is essential. Policy changes may also be required to prevent degradation of water quality. Therefore, the role of the CCTF is to insure that NYCDEP's strategic and capital planning efficiently take into account the potential effects of climate change. In addition, New York City has looked more broadly at the vulnerability of its water supply system to climate change. Major and Goldberg (2001), for example, provide a detailed review of the impacts of global climate change on the New York City Water Supply System, and examine types of adaptation that might be undertaken to cope with climate change.

Water storage

Intensification of the hydrological cycle could make reservoir management more challenging, since there is often a tradeoff between storing water for dry-period use and evacuating reservoirs prior to the onset of the flood season to protect downstream communities. It may become more difficult to meet delivery requirements during prolonged periods between reservoir refilling without also increasing the risk of flooding. Earlier spring runoff from snowmelt is a likely manifestation of global warming. Much of Europe and the western United States depend on snowmelt as a water source for most of the year, so earlier runoff clearly affects water storage on a broad scale. To the extent that adequate reservoir space is available, changes in reservoir management practices could mitigate some of these effects. Seasonal climate forecasts might provide some adaptation leverage for reservoir managers. For example, forecasts based on the current state of the El Nino Southern Oscillation and other large-scale climatic indices correlate well with precipitation patterns in some regions, and would be useful information for reservoir management decisions.

Water demand

Future climate change could influence municipal and industrial water demands, as well as competing agricultural irrigation demands. Municipal demand depends on climate to a certain extent, especially for garden, lawn, and recreational field watering, but rates of use are highly dependent on utility regulations. Shiklomanov (1998) notes different rates of use in different climate zones, although in making comparisons between cities it is difficult to account for variation in non-climatic factors. Studies in the UK (Herrington 1996) suggest that a rise in temperature of about 1.1°C by 2025 would lead to an increase in average per capita domestic demand of approximately 5 percent – in addition to non-climatic trends – but would result in a larger percentage increase in peak demands, since demands for garden watering may be highly concentrated. Boland (1997) estimated the effects of climate change on municipal demand in Washington, D.C. under a range of different water conservation policies and concluded that the effect of climate change is "small" relative to economic development and the effect of different water conservation policies.

Industrial use for processing purposes is relatively insensitive to climate change; it is conditioned by technologies and modes of use. Demands for cooling water would be affected by a warmer climate because increased water temperatures will reduce the efficiency of cooling, perhaps necessitating increased source water withdrawals to meet cooling requirements (or, alternatively, changes in cooling technologies to make them more efficient).

Climate Change Information in Utility Planning and Adaptive Management

As discussed previously, there are several layers of uncertainty inherent in assessing climate change impacts. For example, uncertainties in projected greenhouse gas emissions, limitations of climate models, information loss when climate projections are downscaled to watershed resolution, and imperfections in hydrological models all contribute to the uncertainty. Perhaps even more frustrating is the fact that there is no universally accepted standard for quantifying these uncertainties. This means that it is difficult to define a meaningful confidence level for these projections. Given the uncertain nature of climate impact analysis, it may be tempting to disregard climate change in decision analysis. However, much is known regarding climate change and the remaining uncertainties do not provide a valid excuse to dismiss all aspects of climate change in water resource planning. Rather, the uncertainty introduced by climate emphasizes the importance of incorporating flexibility or no-regrets options in water resource planning.

Water managers are accustomed to adapting to changing circumstances, many of which are analogs of future climate change, and they have developed a wide range of adaptive options. Supply-side options are more familiar to most water managers, but demand-side options are becoming increasingly prevalent. Water management is evolving continually, and this evolution will affect the impact of climate change in practice. For reasons noted above, climate change will inevitably challenge existing water management practices, especially in countries with less experience in incorporating uncertainty into water planning. The current challenge is to incorporate climate change uncertainty along with the other types of uncertainty traditionally treated in water planning.

A review of the scientific and water planning literature suggests that most water resource and water utility studies have incorporated climate change information into their planning process using a top-down approach (Figure 29). This approach typically begins by establishing the scientific credibility of human-caused climate change, develops future climate scenarios to be used at the regional level, and then imposes those potential changes on water resource systems to assess, for example, system reliability. The problems with a top-down approach are: 1) it does not always address the unique vulnerabilities and information needs of a utility, and 2) the approach may become mired in the uncertainty of the future climate projections. There is a danger that utility managers will disregard the results and view them as lacking relevancy and credibility. Alternatively, the bottom-up approach begins by identifying a water utilities' most critical vulnerabilities; articulates the causes for those vulnerabilities; suggests how climate change, climate variability, and climate extremes might or might not exacerbate those vulnerabilities; and finally designs an analytic process to better address and solve the vulnerability in the face of the climatic uncertainty (e.g. a precautionary approach) (Figure 29). In either top-down or bottom-up approaches, Integrated Water Resource Management (IWRM) can be the most effective method for assessing adaptation options and their implications in the context of an evolving regulatory environment with its competing demands (Figure 29).

Integrated Water Resource Management

IWRM (Bogardi and Nachtnebel 1994; Kindler 2000) is a systematic approach to planning and management that considers a range of supply-side and demand-side processes and actions, and incorporates stakeholder participation in decision processes. It also facilitates adaptive management by continually monitoring and reviewing water resource situations. To plan effectively, utilities must engage their customers and external regulators when assessing the potential impacts of climatic change on their water systems. IWRM is a useful tool that utility managers can apply in their efforts to plan for adaptation to climate change.

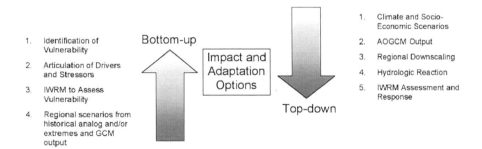

Figure 29. Bottom-up and top-down approaches to climate change assessment.

To articulate the supply- and demand-side processes and actions, IWRM must simultaneously address the two distinct systems that shape the water management landscape. Factors related to the biophysical system, namely climate, topography, land cover, surface water hydrology, groundwater hydrology, soils, water quality, and ecosystems, shape the availability of water and its movement through a watershed. Factors related to the socio-economic management system, driven largely by human demand for water, shape how available water is stored, allocated and delivered within or across watershed boundaries. Increasingly, operational objectives of the installed hydraulic infrastructure constructed as part of the management system seek to balance water for human use and water for environmental needs. In Europe, for example, the EU Water Framework Directive obligates water utilities to cooperate in river basin management efforts to achieve good ecological and water- quality status for rivers and lakes. Thus, integrated analysis of the natural and managed systems is arguably the most useful approach to evaluate management alternatives.

This type of analysis relies upon the use of hydrologic modeling tools that simulate physical processes including precipitation; evapotranspiration, runoff, infiltration, etc. (see Figure 30a, Pre-Development). In managed systems, analysts must also account for the operation of hydraulic structures such as dams and diversions (see Figure 30b, Post-Development), as well as institutional factors that govern the allocation of water between competing demands, including consumptive demand for agricultural or urban water supply or non-consumptive demands for hydropower generation or ecosystem protection. Changes in each of these elements can influence the ultimate impacts of climate change on a water utility and its customers.

There are several approaches to using climate change information from AOGCMs to evaluate the response of the terrestrial hydrologic cycle at scales relevant to water utilities, each differing in the detail used to represent various physical processes. Although different hydrologic models can yield different values in terms of streamflow,

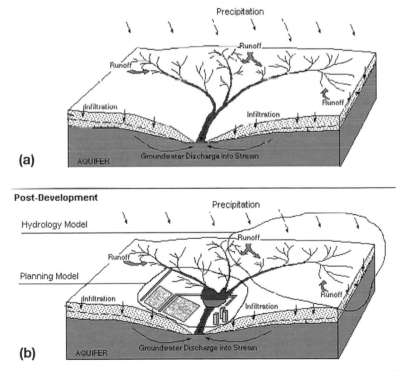

Figure 30. Characterization of (a) pre- and (b) post-watershed development that highlights the implications of water resource infrastructure on the hydrologic cycle.

groundwater recharge, water quality results, etc (Boorman and Sefton 1997; Beven 2001), their differences have historically been small in comparison to the uncertainties attributed to climate change reflected in the differences among AOGCM output. However, the chain of effects from climate, to hydrologic response, to water resource systems, to the actual impacts on water supply, power generation, navigation, water quality, etc. will depend on many factors, each with a different level of uncertainty.

Deciding how to evaluate system reliability or specific vulnerabilities given future uncertainty is a major challenge for water resource managers. Note that both the top-down or bottom-up approaches ultimately require climate and socio-economic projections for assessing particular vulnerabilities (bottom-up) or overall performance (top-down). While current water resource planning methods already consider projected changes in water demand and variations in water supply, they often rely upon limited historical datasets. Historically, many water utilities made their infrastructure investments and long-term management strategies assuming that precipitation and runoff would follow past trends. Mounting evidence for climate change makes this an increasingly tenuous assumption.

> *Deciding how to evaluate system reliability or specific vulnerabilities given future uncertainty is a major challenge for water resource managers.*

Assessing vulnerability to climate change

A water utility will need to develop scenarios for use in assessing the role of climate, as it investigates potential climate change impacts on the broader water resource system (top-down) and investigates particular vulnerabilities (bottom-up). Selection and application of baseline and scenario data occupy central roles in most climate change impact and adaptation analyses. These can include the development of alternative time series datasets of important meteorological variables, such as daily or monthly precipitation, temperature, wind speed, humidity etc.; projections of land use and land cover change; and socio-economic scenarios of population change and water use rates for demand modeling; and others. Detailing the different methods and approaches is beyond the scope of this Primer, but we offer a brief review and suggestions on some of the more straightforward ways to develop climate change scenarios for assessing vulnerabilities. The National Center for Atmospheric Research and Awwa Research Foundation plan a future partnership project to promote the development of analysis tools for evaluating vulnerabilities to climate change.

Downscaling

Downscaling is an attempt to understand how local scale processes might respond to larger-scale weather and climate changes as represented by AOGCMs, with downscaled datasets sometimes used to force hydrology and land-surface related models. A key reason for developing downscaled datasets is the hypothesis that the statistical characteristics of the downscaled data will contain differences relative to the historic record that reflect regional climatic changes – not only in the mean condition but also in other statistical attributes such as the sequences of storms and dry periods.

Climate scenarios generated through downscaling techniques involve the development of statistical relationships between historic meteorological observations and outputs from regional and/or global climate models. These methods include: 1) adjusting historical temperature observations by adding a fixed temperature increment to the historical record and multiplying historical observations of precipitation by a fixed amount, with the fixed amount sampled from the AOGCMs; 2) statistical methods guided by AOGCM output; 3) trained stochastic models based on the relationship between GCM atmospheric circulation patterns and surface variables; 4) statistical-dynamical downscaling which utilizes regional weather models driven by coarser-scale GCM boundary and initial conditions to resolve the finer-scale atmospheric processes. These processes are then related to surface variables. In the case of statistical-dynamical downscaling, the fine scale processes are then related to surface variables. This method captures the stochastic characteristics of large area circulation patterns, which are arguably better represented by GCMS than are surface processes, most notably precipitation. Wilby et al. (2004) provides a useful guide to statistical downscaling methods.

While these approaches for generating climate scenarios for impact analysis are useful, they do have limitations. For example, a climate change scenario might be generated from large-scale AOGCM features such as pressure patterns that are then related to storm tracks and thus surface precipitation. However, the AOGCM's simulation of past and current climate might show biases in the current pressure patterns, which would then be propagated to the downscaled precipitation estimate. Thus, this technique in producing either current or future climate sequences for impact analysis can be problematic. So, if an AOGCM is so biased that it does not adequately replicate the historic climate of a region, the level of confidence placed in its ability to generate scenarios of the impacts due to increasing CO_2 and aerosol changes diminishes.

Stochastic weather generators have also been used to develop climate datasets for impact analysis. These can address some of the issues just raised with their ability to simulate plausible climate scenarios, and have themselves been used as downscaling techniques in global change studies (Wilks 1992). Typically, a stochastic weather generator is developed based on the historically observed data at a location, and can then be used to simulate climate scenarios consistent with the global change scenarios. However, Katz (1996) points out that modifying the parameters of a stochastic model can lead to unanticipated effects. For example, modifying the probability of daily precipitation occurrence using a stochastic weather generator (Richardson 1981) also changes the mean and standard deviation of the daily temperature as well.

Incremental and analogue

Doing such downscaled analyses based on the output of multiple climate model simulations can be a very laborious and time-consuming task. The daunting prospect of developing detailed climate datasets for impact analysis has led a number of researchers to use simpler, almost back-of-the-envelope approaches to explore the possible implications of climate change for water resources. Several analyses have used hypothetical changes in temperature and precipitation amounts by simply scaling a historic record by some predefined amount, essentially amounting to a sensitivity analysis to a climate perturbation. A drawback of this approach is that the hypothetical scenarios may not be internally consistent (e.g. increases in temperature might correspond to decreases in precipitation, while such a hypothetical scenario might not capture the climatic reality). Despite those drawbacks, systematic analysis of such scenarios can be useful for delineating the relative importance of changes in temperature and precipitation and can provide an inexpensive way to explore vulnerabilities of water supply systems, water quality, and in-stream resources.

Another approach to identify potential worst-case climate scenarios on a regional or local scale involves using data from historical extreme events, such as a region's most severe drought in the past century or climate traces developed from tree-ring studies. This approach has the advantage of realism because events that occurred in the past could occur again. The major disadvantage is that it makes no attempt to account for the effects of global climate change. Of these various approaches, no one method is categorically superior, but rather specific hydrological characteristics of the watershed of interest should determine which technique is appropriate.

Several utilities such as Portland, Seattle, and many utilities in England, have used downscaling techniques to derive alternative climate data for use in hydrological models. The variety of climate models and hypothetical parameters used to develop a range of possible climate variation influence the robustness of this method. In general, it is wise to consider several climate model simulations to get a sense of the range of possible changes.

Case Study: Options for Assessment – Boulder, CO

The City of Boulder, Colorado completed a study that evaluated 12 potential water supply/demand 'futures' for the city, including four alternative projected future water demands with three hypothetical climate scenarios. The intent of the study was to evaluate the long-term adequacy of the city's water supply system. The study made use of a 300-year tree-ring hydrologic reconstruction to derive alternative hydrologic traces based on changes in mean flow and annual variability. The study took a sensitivity approach to investigate the vulnerability of their system to climate variability. Climate change studies provided the bounds for their stylized scenarios, which were constructed to define reasonable worst-case outcomes:

> While some research has suggested that climate change may result in earlier runoff and lower late summer stream flows due to more rain and less snow, we did not attempt to redistribute seasonal stream flows in this scenario. This decision was made for the sake of simplicity and conservatism. The degree of shift in seasonal runoff patterns has not been suggested by research to date. Earlier runoff is likely to increase the yield of Boulder's water supply because Boulder's reservoirs would be able to store more water before the onset of the irrigation season (Hydrosphere Resource Consultants 2003, p. 3, used with permission of the City of Boulder, CO USA).

Based on the scenarios examined, the study concluded that if climate change results in significantly reduced streamflows, Boulder's water supply system would not be able to meet future water demands in some drought years with a reasonable margin of safety. The study projected that Boulder will be able to meet future water needs up to a defined level of reliability in 3 out of 4 of the water demand scenarios under present hydrologic conditions and one of the scenarios assuming streamflow changes. However, in 7 of the 8 scenarios assuming large streamflow reductions due to global climate change, it was necessary to implement water use restrictions more frequently than allowed under the city's presently accepted reliability criteria. If streamflow variability increased by 25 percent, Boulder is projected to be able to meet future water needs by applying drought year water use restrictions slightly more often than presently anticipated. However, actual shortages in supply occurred in some drought years under scenarios assuming a 15 percent reduction in streamflow even with a greatly increased application of water use restrictions. This would be the case unless additional water supplies are acquired or developed, or additional reductions in per capita water use are achieved beyond the levels anticipated in Boulder's comprehensive water conservation program.

Case Study: Using Wildfire Experience to Assess and Mitigate Vulnerabilities – Denver Water, CO

A number of studies suggest that wildfires could become more frequent in many regions as climate change imposes new stresses on vegetation. Vegetation patterns will change, through time, in response to the changing climate, and fire will likely play a role in that evolution.

Denver Water has experienced substantial water quality impacts due to wildfire. In 2002, in the midst of a severe drought, the largest wildfire in the history of Colorado literally surrounded Denver Water's Cheesman Reservoir, burning 97 percent of the 7,245 acres owned by the utility. That fire, known as the Hayman fire, burned a total of 138,000 acres in the Denver Water collection area – 2/3 of that above Cheesman

Reservoir. The utility has incurred substantial costs following the Hayman burn for post fire clean-up, erosion control and management of water quality degradation. Denver Water has spent $6.5 million to date on the fire. That, however, pales in comparison to the impacts of a much smaller fire that occurred six years earlier.

The Buffalo Creek fire burned 11,900 acres on May 18,1996. The intense fire burned directly along the drainage of a seemingly insignificant tributary to the upper South Platte River. The upper South Platte is a major source of Denver's water supply, but Denver Water also collects water from other watersheds – notably the upper Colorado on the western side of the Continental Divide. While Buffalo Creek itself contributes only a trivial share of Denver's water supply, it is strategically located directly upstream of the critically important Strontia Springs Reservoir which is the intake point for the Foothills Treatment Plant. The Strontia Springs/Foothills facilities typically handle approximately 80 percent of Denver's water, collecting inflow from both the upper South Platte – through Cheesman Reservoir – and from Colorado's Western Slope – through the Roberts Tunnel.

Two months after the Buffalo Creek fire, heavy thunderstorms directly over the denuded burn area resulted in a flash flood that killed 2 people and washed tons of sediment and debris down the creek and into Strontia Springs Reservoir. On July 12, 1996 more sediment washed into the reservoir than had accumulated over the course of the previous thirteen years. In that single day, the 770 acre-foot reservoir lost an estimated 30 years of its planned 50-year useful life. Ash, charred debris and litter also washed into the reservoir and threatened to clog the intakes to the treatment plant. The debris flow necessitated emergency cleanup operations costing nearly $1 million. Those immediate cleanup costs were only the tip of the iceberg because elevated turbidity has become a chronic problem in the reservoir, requiring an additional $250,000 in water treatment costs each year. In addition, dredging costs necessitated by the rapid sedimentation, may amount to $15–20 million dollars over the next 10 years.

Long-term impacts of both fires include ongoing erosion, sedimentation and transport of heavy metals and micronutrients into the reservoirs. Algae blooms, leading to sharp reductions in dissolved oxygen levels in the immediate post-fire period, and longer-term growth of aquatic weeds, including milfoil have been some of the unanticipated consequences of the increased nutrient supplies since the fires. The utility is now working actively with federal, state and private partners to restore the watershed and to improve land use and vegetation management to reduce the potential for future catastrophic fires, and to increase the resilience of the watershed to the possible effects of climate change.

A lesson from Denver's experience is that the significance of wildfire impacts on water supplies partially depends on the vulnerability of the utility's facilities, and not solely on the size of the fire. In retrospect, officials from Denver Water note that the Buffalo Creek fire might have had less severe consequences if greater attention had been given to mitigating the fuel load buildup in that watershed prior to the fire event. In addition, they note that had they installed sensors upstream of the Strontia Springs reservoir to monitor the pulse of debris and sediment coming down the river, the utility could have shut down its treatment plant earlier, thus preventing some of the damage. Finally, the utility's experience suggests that water utilities should be proactive in assessing the potential consequences of wildfires, particularly with respect to sediment and debris flows in determining appropriate rapid responses during actual crisis events.

Infrastructure – evaluating options in the context of climate change

Construction of new reservoirs is a possible way to prepare for longer dry periods, but there are drawbacks to large infrastructure projects. There are a limited number of good sites remaining, and the economic, environmental, and social costs associated with reservoir construction can be prohibitive. Even in locations where storage capacity is high relative to construction costs, the ecological damages of diverting water from streams can be prohibitively costly from a social standpoint. Increasing storage capacity of existing reservoirs is one option, but increasing the surface area as climate warms would potentially increase evaporative losses. To justify the cost of infrastructure projects, reservoirs must be functional for many years. Their expected value is more difficult to predict with the possibility of climate change. An increase in runoff could make additional storage superfluous, while if net inflows decrease, the additional capacity might go unused. Therefore, utilities must consider possible changes in future hydrology when assessing the long-run benefits of storage infrastructure. Despite these complexities, there are regions where expansion of surface storage appears to be the best option for meeting future water needs. As demonstrated in the Portland case study at the beginning of this document, infrastructure that will be cost-effective in a broad range of possible futures is one example of a sensible adaptive strategy.

In addition to developing new infrastructure, it may be possible to re-operate existing infrastructure to counter potentially negative climate change impacts. Given contemporary operating procedures, a number of studies have demonstrated that the reliability of reservoir yields can vary dramatically with only a small change in reservoir inflows (Nemec and Schaake 1982; Lettenmaier and Sheer 1991; McMahon et al. 1989; Mimikou et al. 1991; Nash and Gleick 1991, 1993). For example, a change in the timing of peak runoff could reduce the likelihood of reservoir refill if current flood control rule curves remain unchanged. One can question whether reservoir-operating rules could be adapted to maintain deliveries with current infrastructure given a change in inflows. Lettenmaier and Sheer (1991) demonstrated that this is indeed possible, but perhaps at the cost of increasing flood risk. Utilities can use sensitivity analyses to examine the effects of changing their operating procedures in light of possible hydrological intensification and the impacts of changes in snowmelt patterns on water storage and flow timing.

Water market options, transfers and water banks

In regions where there are limited options for increasing supplies, utilities have begun to negotiate with competing water users, most often in agriculture, to acquire additional water. In most parts of the world, the agricultural sector dominates water withdrawals and consumptive water use. For example, in Oregon, irrigation accounts for about ninety-six percent of consumptive use (Solley et al. 1998), and agriculture is the dominant water user throughout the western United States. Therefore, a relatively small transfer from agriculture can add substantially to a municipal water supply. Often, such transfers can benefit both urban water users and farmers. When the value of water for crop production is lower than the price that a utility is willing to pay for it, it is financially advantageous for a farmer to sell water. However, there are often a number of legal or institutional barriers that inhibit transactions between farmers and municipalities. This section focuses on the success utilities have had acquiring water through permanent transfers, dry-year contracts, spot markets and water banking.

For utilities seeking to increase their withdrawals on a consistent basis, purchasing permanent water rights from farmers can be the most cost-effective option. However, the negotiation process is time-consuming and often entails high transaction costs due to legal complexities and reluctance of farmers to relinquish water rights permanently. In addition, utilities may only need additional withdrawals during dry years, making permanent acquisition of new supply superfluous in other years. These obstacles have led many utilities to precede or replace negotiations for permanent water transfers with temporary trading of water rights. Utilities have successfully employed several types of temporary transfers, including dry-year option contracts and spot market purchases.

A dry-year option contract is made between a utility and an individual farmer or agricultural district. Within the term length of the contract, the utility has the option during dry years to withdraw water that is usually dedicated to agriculture. Typically, the contract is tied to reservoir levels, but the timing of the transfer and the term length of the contract are also key issues that must be covered. The utilities pay a fee to secure the water option, and if the option is exercised they are required to compensate the farmers for lost crop revenue, disruption of farm planning, and other similar costs. As will be seen in the following case study, the revenue that farmers generally receive from utilities is considerably larger than the value of the water from crop revenue, and water transfers do not exclude them from growing low water-intensive crops. Therefore, dry-year contracts are often attractive to farmers because they are only temporary agreements and they tend to have low transaction costs. While they are more expensive than permanent water transfers for meeting long-term municipal needs, dry-year contracts are a good way to introduce water trading into a region. Once a region is familiar with the terms and outcomes of water transfers, the cost of negotiating permanent transactions can be reduced significantly.

Spot markets are another option for utilities seeking temporary water transfers when supplies are low. This type of transaction is a one-time lease for a specific quantity of water. The transfer cost for spot markets is low and the negotiations are brief, since the agreement is made with individual water users and does not go through government water agencies. A useful application of spot markets is that they provide a last-minute supply option during unforeseen shortages; however, the later the negotiation the more expensive is the acquisition. In general, planning in advance will dramatically decrease the cost of water transfers.

Utilities have used several transaction methods to secure water transfers. The most common are acquisitions negotiated case by case, which have the advantage that they are tailored to meet the specific needs of a utility. However, such transactions tend to have high transaction costs, and the public may regard them as secret deals that may be unfair to farmers who do not have the opportunity to sell water. Another commonly used option is a standing offer, where the utility fixes and publishes the price it is willing to pay for water. The transaction costs are low for this straightforward arrangement, but it is often difficult to set an appropriate offer price. Because the utility must commit itself to an initial price, it is easy for municipalities to acquire too little volume or be obligated to pay for more water than they can use. A third method, the use of bidding mechanisms to secure transfers, is becoming increasingly common in the United States. This method has the advantage of being transparent, and therefore having a public perception of fairness. One difficulty with this system is that many areas have an initial lack of familiarity with bidding systems. Also, auctions must have clear rules and procedures, and must be carefully designed to prevent water sellers from raising the price of water by collusion. However, a well-designed bidding mechanism will reflect current market conditions and result in an efficient method for acquiring supply.

Regional water banks can facilitate water transactions between municipalities and other sectors. Water banks reduce transaction costs by coordinating negotiations between buyers, and standardizing trading procedures. The banks typically use water stored in aquifers or reservoirs to bring about temporary transfers for dry year needs. Therefore, they work well in regions where there is adequate reservoir space or other storage options exist.

Case Study: Adaptive Management – Metropolitan Water District, CA

The Metropolitan Water District of Southern California (Metropolitan) is a wholesale water supplier for utilities in Southern California, including Los Angeles, Orange, San Diego, Riverside, San Bernardino and Ventura counties. Southern California depends upon the Colorado River and the State Water Project (SWP) – sources originating outside of Metropolitan's service area – to meet roughly half of its retail water demands. In anticipation of future population growth and other factors that might affect the reliability of imported supplies, Metropolitan approved its Integrated Resource Plan (IRP) in 1996 to develop new and diverse approaches to improve supply reliability. Over the past decade, imported water supplies have been complemented by aggressive conservation programs, development of local water recycling and groundwater supplies, enhanced water storage and conveyance, and water transfers. In July 2004, Metropolitan updated its IRP to continue adapting to changes in its diverse supply portfolio, notable examples being recent reductions in surplus deliveries from the Colorado River and threats of levee failures in the Sacramento-San Joaquin Delta which could imperil SWP supplies). This integrated resource approach has been deemed successful by Metropolitan and is characteristic of what Metropolitan views as a "no regrets" approach to managing demands in its service area. By balancing demand management efforts with development of new storage and transfer programs, Metropolitan expects to continue to provide reliable water supplies to Southern California.

The use of program- and device-based conservation to help avoid reliability issues associated with severe drought has increased steadily since the IRP was initiated. While these programs reduce water use in all years, during dry years, the value of such programs is more visible as reduced consumption allows water managers more flexibility with various resources and keeps demands for imported supplies within certain levels even with a growing population. Since implementing the IRP, the use of local resources in Metropolitan's service area has increased significantly through the development of water recycling projects, groundwater recovery, and increased groundwater and surface storage. In 2004, nearly one million acre-feet of water were produced through these local programs to meet Southern California water supply needs.

In recent years, Metropolitan has helped develop more than 75 water recycling and groundwater recovery programs with local agencies through funding incentives. One example is the West Basin Water Reclamation Program. The West Basin Municipal Water District receives treated wastewater from the City of Los Angeles (this water has undergone a level of recycling called secondary treatment), treats it further (tertiary treatment) and delivers the water primarily for landscape irrigation and various industrial purposes. A portion of this extensively treated water undergoes further purification (reverse osmosis) and is injected into the ground to maintain a barrier against seawater intruding into drinking-water wells in the South Bay area. This project currently produces over 20,000 acre-feet of water each year (AFY), and is expected to expand production to around 70,000 AFY by 2025 to help meet local demands.

Metropolitan recently selected 13 new projects through a competitive Request for Proposals process that concluded in April 2003. These projects, which include both recycling and groundwater recovery programs, are projected to provide approximately 65,000 AFY to meet future demands within Metropolitan's service area. One project with the Municipal Water District of Orange County and the Orange County Water District will provide 31,000 AFY for injection into the ground to provide a seawater intrusion barrier to protect supply wells in Orange County. These programs provide reliability and flexibility to water managers in the region, and help reduce demand for imported water supplies.

In the past decade, Metropolitan has also increased its storage capacity tenfold through completion of the Diamond Valley Lake in Hemet, new groundwater storage programs inside and outside of its service area, and by acquiring contractual storage in state reservoirs. The expansion in storage capacity has reduced the risk of water shortages in a single year, and also allows for water purchased from other entities, such as transfers from the agricultural sector, to be reliably managed.

Metropolitan has been a leader in the use of voluntary water transfers to improve supply reliability. California has a large agricultural sector, and Metropolitan's distribution system is linked to agricultural districts through other water conveyance systems throughout the state. Increasingly, voluntary water transfers are becoming part of an overall business strategy for agricultural interests who are recognizing that in some years it may be more profitable to sell their water supplies than grow crops. These water transfers generally provide a level of revenue certainty for farmers and are typically structured to provide regional benefits as well. Water utilities benefit from the increased flexibility afforded by additional source of supplies. When all potential risks to production are considered, there are times when farmers find it advantageous to sell their water, at rates that are both good for them and economical for utilities. The revenue from the sales provides a means for farmers to maintain viable agricultural operations.

An example of a temporary voluntary transfer from agriculture to municipalities occurred in 2003. Anticipating reduced SWP supplies, Metropolitan signed option contracts with several agricultural districts in the Sacramento Valley. A flat fee of $10 per acre-foot of water secured the option, and farmers received $90 per acre-foot for each option Metropolitan exercised. Metropolitan secured options for approximately 150,000 acre-feet, of which approximately 100,000 acre-feet were exercised. Transfer programs such as this are a cost-effective way to increase reliability, and Metropolitan plans to continue working with its agricultural partners to develop such agreements for future years.

As an example of a long-term voluntary water transfer, in August 2004, Metropolitan and the Palo Verde Irrigation District (PVID) executed a 35-year agreement to implement their Land Management, Crop Rotation and Water Supply Program. Under the agreement, individual landowners will agree not to irrigate up to 29 percent of the valley's farm land at Metropolitan's request, thereby creating a water supply of up to 111,000 acre-feet for Metropolitan. In addition to boosting Metropolitan's water reliability, the program is also designed to stabilize the Palo Verde Valley economy. Like the pilot Metropolitan/PVID program effort that took place from 1992 to 1994, the farmland can remain as prime agricultural acreage and will be neither "retired," nor converted to another use. Landowners will receive a one-time payment per acre allocated, and additional annual payments for each acre not irrigated under the program in that year. To offset possible adverse impacts on the community, Metropolitan has authorized an estimated $6 million for local community improvement programs.

While research into potential effects on water supplies from climate change continues to grow, there are still few if any certainties. Metropolitan believes that its integrated resource approach will enable it to continue to manage demands in its service area. A diverse portfolio of supplies that includes development of new storage and transfer programs characterizes Metropolitan's "no regrets" approach to dealing with uncertainty while continuing to provide reliable water supplies to Southern California.

Demand management

In regions where expansion of supply infrastructure is infeasible, demand management is a sensible strategy to meet future water needs. Throughout most of the 20th century, the United States met increasing demand for water principally by expanding water withdrawals, but by the end of the 1970s, as expanding withdrawals became constrained by high costs and environmental regulations, it became apparent that new strategies were needed to deal with increasing demand. Improvements in water use efficiency have been encouraged by price incentives, water transfers, improvements in technology, and regulations. Demand management strategies have become increasingly important in satisfying the United States' changing needs for freshwater. Non-structural solutions such as these will be important in confronting impacts of climate change as well.

There are two main components of non-structural water resource management: improving the efficiency of water use and effective reallocation of saved water. Water use efficiency can be improved through technology, economics and institutions. Currently large water losses are incurred because of leakage from aging distribution systems to the groundwater system; in California 10 percent of water supplied for urban use is lost in distribution. Improvements in the distribution infrastructure can alleviate supply shortages while avoiding the environmental and social costs associated with increasing water storage infrastructure. In addition, metering and price structures that encourage conservation are important demand management tools.

Conclusion

Climate is one of many sources of uncertainty affecting water utilities. Some would argue that the impacts of climate change are so uncertain and so far in the future that they pale in significance when compared to more immediate concerns. However, water utilities should not ignore this risk, because this new source of uncertainty is significant for long-term planning. Planning for climate change will improve resilience to droughts and floods that arise from ongoing climate variability. The incomplete nature of our understanding of the local effects of global climate change raises different challenges for water management than municipal water providers routinely face when dealing with normal climate variability.

To plan effectively for the future, utilities should assess the potential impacts of a range of plausible climate change scenarios on their ability to meet customer needs and comply with quality standards and environmental objectives in a cost-effective manner. This requires rethinking traditional approaches to the planning process that rely on assumptions such as climate stationarity. Scenarios based on climate model output are a tool that utilities can use for these assessments, but it is important to understand that no single climate model can yield a reliable projection of future climatic conditions. If climate model output is used, it must be appropriately downscaled to the relevant watershed level, and any analysis should use projections from several models to generate a range of physically plausible scenarios of the impacts of climate change on the utility's water resources. The utility can then use the resulting hydrologic projections to evaluate the robustness of alternative response strategies given the unavoidable uncertainties arising from climate change.

The previous chapters of this Primer have emphasized that using climate model output in a top-down approach to assessing climate-change impacts is not the only way to tackle this problem. Valid, serious assessments can take the bottom-up approach of beginning with an assessment of the utility's vulnerabilities. Regardless of which approach a utility selects as the most appropriate for its own situation, it is clear that all utilities can benefit by learning from the experience of others in the industry and by becoming well-informed about the science of climate change. Future research sponsored by Awwa Research Foundation and the National Center for Atmospheric Research will enhance utility efforts to evaluate and plan for the effects of global climate change.

Glossary

Acre-foot – A volume of water equal to 325,900 gallons (or 1.233 million liters).

Anthropogenic – Human-induced.

AOGCM – see Coupled Atmospheric-Ocean General Circulation Model

Aphaelion – The point in the earth's revolution when the earth is farthest from the sun.

Climate change – Any trend or persistent change in the statistical distribution of climate variables (temperature, humidity, wind speed, etc.)

Conjunctive Use – A term used broadly to define any strategic combined use of surface water and groundwater, usually emphasizing the use of surface water during wet periods and groundwater reserves during dry periods.

Coupled Atmospheric-Ocean General Circulation Model (AOGCM) – This is a type of General Circulation Model (see definition) where a mathematical representation of the atmosphere is coupled to models of other components of the climate system, such as oceans and sea ice. Coupling the atmosphere to other climate components provides a more realistic model of climate change.

Downscaling – Procedures for translating data from General Circulation Models (GCMs) (and other tools generating output at large geographic scales) to small geographic scales, such as individual watersheds.

Draw-down processes – In the case of greenhouse gases this refers to natural processes that remove these gases from the atmosphere, such as uptake of CO_2 in biota, soils or oceans.

ESA – Endangered Species Act. Federal legislation (1973 as amended) prohibiting actions that kill, harm, or otherwise harass members of species recognized as endangered.

El Niño – A large scale warming of the eastern tropical Pacific Ocean. In the American West, El Niño generally brings increased precipitation to the Southwest and reduced precipitation to the Northwest.

El Niño/Southern Oscillation (ENSO) – A general term used to describe both warm (El Niño) and cool (La Niña) ocean-atmosphere events in the tropical Pacific as well as the Southern Oscillation, the atmospheric component of these phenomena.

Earth energy budget – Is determined by the balance between the amount of energy coming from the sun and the energy radiated back into space in the form of infrared radiation. If this balance is upset by, for example, a change in the amount of solar radiation reaching Earth, or by a change in the amount of greenhouse gases in the atmosphere, then Earth's surface either warms or cools until a new balance is established.

Ensemble Technique – The production of findings or projections by using a collection of model runs, rather than relying exclusively on the output of any one run (or ensemble member).

ENSO – See: El Niño/Southern Oscillation.

Feedback – An interaction in which a change in one process triggers changes in a second process, which then cause further changes in the initial process. A positive feedback intensifies the initial effect and a negative one reduces it.

GCM – See: General Circulation Models. [Also occasionally defined as "Global Climate Model"]

General Circulation Models (GCM) – Sophisticated mathematical computer models of the atmosphere and its phenomena over the entire Earth, based on equations of motion and considering radiation, photochemistry, and the transfer of heat, water vapor, and momentum.

Greenhouse Effect – Process whereby energy from the sun is trapped by certain (greenhouse) gases in the atmosphere (i.e., water vapor, carbon dioxide, nitrous oxide, and methane). Human-induced increases in greenhouse gas emissions are thought to be enhancing this natural phenomenon.

Hydrograph – A graph that illustrates hydrologic measurements over a period of time, such as water level, discharge, or velocity.

Instream Flow Rights – A type of water right administered within the prior appropriation system that calls for water to be left in the stream channel for use, including for environmental purposes.

Intergovernmental Panel on Climate Change (IPCC) – Established by the World Meteorological Organization (WMO) and the United Nations Environment Programme (UNEP) in 1988, the role of the IPCC is to assess scientific, technical and socio-economic information relevant to global climate change.

Intertropical Convergence Zone (ITCZ) – A broad band that girdles the equator and is characterized by rising air, frequent convective storms, and high annual precipitation.

Jet streams – Broad wind bands that flow from west to east. They are largely responsible for storm movements in temperate regions.

IPCC – See: Intergovernmental Panel on Climate Change.

La Niña – A large scale cooling of the eastern tropical Pacific Ocean occurring at irregular intervals of between about two and seven years and lasting for one to three years. In the American West, La Niña generally brings increased precipitation to the Northwest and reduced precipitation to the Southwest.

Maf – million acre-feet

Milankovich cycle – The cycle of change in the angle of tilt of the Earth's axis, which varies over a 41,000 year cycle.

NAO – see North Atlantic Oscillation

North Atlantic Oscillation (NAO) – A long-term ocean temperature fluctuation of the Atlantic basin, similar to the Pacific Decadal Oscillation. This oscillation measures swings in the relative intensity of the winter low-pressure cell centered over Iceland and the high-pressure cell centered over the Azores.

Pacific Decadal Oscillation (PDO) – A long-term ocean temperature fluctuation of the Pacific Ocean. The PDO waxes and wanes approximately every 20 to 30 years.

PDO – See: Pacific Decadal Oscillation.

Perihelion – The point in the earth's revolution when the earth is closest to the sun.

polder – A tract of lowland reclaimed from the sea by dikes, dams, etc.

ppbv – parts per billion

ppmv – parts per million

Recharge – The movement of water from the Earth's surface into aquifers, either through natural processes or active management.

Stochastic weather generator – Historically observed data at a location is used to simulate local climate scenarios consistent with global change scenarios. This provides a hypothetical climate dataset for impact analysis in a particular region.

SWE (or SWC) – Snow Water Equivalent (or Content). A measure of how much water is contained within a given snowpack.

Thermohaline circulation – The connection between the movement of cold, salty water in the ocean's depths and the movement of warm, less saline water at the surface.

Water Bank – There are two types of water banks: 1) groundwater storage and recovery programs; 2) formal mechanisms created to facilitate voluntary transfers of water from owners of existing water rights to other users.

Water Right – A legally recognized and protected privilege conferred upon individuals and organizations to use water under given terms. Water rights in the US West are generally based on the prior appropriation doctrine, and are defined in terms of the quantity, location, timing, purpose, and seniority of the water use.

Water Transfers – The voluntary movement of water and water rights between sectors and regions through the use of markets. In the US West, most transfers move water from agricultural to municipal uses.

Index

References

Allen, M.R., and W. J. Ingram. 2002. Constraints on Future Changes in Climate and the Hydrologic Cycle. *Nature,* 51(9):735–751.

Alverson, K.D., R.S. Bradley and T.F. Pedersen (eds.). 2003. *Paleoclimate, Global Change and the Future.* Springer-Verlag, Berlin, Germany.

Arnell, N., B. Bates, H. Lang, J. Magnuson, and P. Mulholland. 1996. Hydrology and Freshwater Ecology. In *Climate Change 1995: Impacts, Adaptations and Mitigation of Climate Change: Scientific-Technical Analyses,* Contribution of Working Group II to the Second Assessment Report of the Intergovernmental Panel on Climate Change. Cambridge, UK: Cambridge University Press.

Arnell, N.W. 2003. Effects of IPCC SRES Emissions Scenarios on River Runoff: A Global Perspective. *Hydrology and Earth System Sciences,* 7:619–641.

Beven, K. 2001. *Rainfall-Runoff Modeling – The Primer.* Chichester, UK: Wiley Interscience.

Blasing, T.J., and S. Jones. 2003: *Current Greenhouse Gas Concentrations.* Carbon Dioxide Information and Analysis Center (CDIAC). Available: <http://cdiac.esd.ornl.gov/pns/current_ghg.html> [cited 12 December 2004]

Bogardi, J.J., and H.P. Nachtnebel. 1994. *Multicriteria Decision Analysis in Water Resources Management.* Report No. WS14. Paris, France: International Hydrological Programme, UNESCO.

Boland, J.J. 1997. Assessing Urban Water Use and the Role of Water Conservation Measures under Climate Uncertainty. *Climatic Change,* 37(1):157–176.

Boorman, D., and C. Sefton. 1997. Recognizing the Uncertainty in the Quantification of the Effects of Climate Change on Hydrological Response. *Climatic Change.* 35(4): 15–434.

Broecker, W.S. 1997. Thermohaline Circulation, the Achilles Heel of our Climate System: Will Man-made CO_2 Upset the Current Balance? *Science,* 278:1582–1588.

Bureau of Meteorology. 1992. *The Greenhouse Effect and Climate Change.* Melbourne: Bureau of Meteorology.

Carter, T.R., M. Hulme, and D. Viner (eds.). 1999. *Representing Uncertainty in Climate Change Scenarios and Impacts Studies.* Report No. 1, ECLAT-2 Workshop. Norwich, U.K., Climatic Research Unit, University of East Anglia.

Cayan, D.R., S. Kammerdiener, M.D. Dettinger, J.M. Caprio, and D.H. Peterson. 2001. Changes in the Onset of Spring in the Western United States. *Bulletin American Meteorological Society,* 82:399–415.

Dettinger, M., and D. Cayan. 1995. Large-Scale Atmospheric Forcing of Recent Trends Toward Early Snowmelt in California. *Journal of Climate,* 8:606–623.

Gates, W.L., A. Henderson-Sellers, G. Boer, C. Folland, A. Kitoh, B. McAvaney, F. Semazzi, N. Smith, A. Weaver, and Q. Zeng. 1996. Climate Models – Evaluation. In *Climate Change 1995 - The Science of Climate Change*. Contribution of Working Group I to the Second Assessment Report of the Intergovernmental Panel on Climate Change. Cambridge, UK: Cambridge University Press.

Giorgi, F., and L.O. Mearns. 1991. Approaches to the Simulation of Regional Climate Change: A Review. *Reviews of Geophysics*, 29(2):191–216.

Grotch, S.L., and M.C. MacCracken. 1991. The Use of General Circulation Models to Predict Regional Climatic Change. *Journal of Climate*, 4:286–303.

Henderson-Sellers, A., and K. McGuffie. 1987. *A Climate Modelling Primer*. Chicester, U.K.: John Wiley.

Hennessy, K.J., J.M. Gregory, and J. F.B. Mitchell. 1997. Changes in Daily Precipitation under Enhanced Greenhouse Conditions. *Climate Dyn.*, 13:667–680.

Herrington, P. 1996. *Climate Change and the Demand for Water*. London, U.K.: HMSO (Her Majesty's Stationery Office).

Hopkinson, C., and Y. Young. 1998. The Effect of Glacial Wastage on the Flow of the Bow River at Calgary, 1951–1993. *Hyrological Processes*, 12:1745–1766.

Hughes, J.P., D.P. Lettenmaier, and E.F. Wood. 1993. An Approach for Assessing the Sensitivity of Floods to Regional Climate Change. In *The World at Risk: Natural Hazards and Climate Change*. Conference Proceedings. New York: American Institute of Physics.

Hydrosphere Resource Consultants. 2003. Simulation of Hypothetical Climate Change Scenarios Using the Boulder Creek Watershed Model, Report to the City of Boulder, CO, September 9.

IGBP (International Geosphere Biosphere Program). 2001. *Global Change and the Earth System: A Planet under Pressure*. IGBP Science No. 4. Stockholm, Sweden: IGBP.

IPCC (Intergovernmental Panel on Climate Change), WG I (Working Group I). 2001. *Climate Change 2001: The Scientific Basis*. Contribution of Working Group I to the Third Assessment Report of the IPCC. Cambridge, UK: Cambridge University Press.

IPCC (Intergovernmental Panel on Climate Change), WGII (Working Group II). 2001. Climate Change 2001: Impacts Adaptation and Vulnerability. Intergovernmental Panel on Climate Change (IPCC), Working Group II Third Assessment Report. Cambridge, UK: Cambridge University Press.

Katz, R. 1996. Use of Conditional Stochastic Models to Generate Climate Change Scenarios. *Climatic Change*, 32:237–255.

Kiehl, J., and K. Trenberth. 1997. Earth's Annual Global Mean Energy Budget. *Bulletin of the American Meteorology Society*, 78:197–208.

Kindler, J. 2000. Integrated Water Resources Management: The Meanders. *Water International*, 25:312–319.

Lettenmaier, D.P., and D.P. Sheer. 1991. Climate Sensitivity of California Water Resources. *Journal of Water Resources Planning and Management*, 117:108–125.

Levitus, S., J.I. Antonov, , T.P. Boyer, and C. Stephens. 2000. Warming of the World Ocean. *Science*, 287:2225–2229.

Major, D.C. and R. Goldberg. 2001. "Water Supply" Chapter 6 in C. Rosenzweig, and W. D. Solecki, (eds.) *Climate Change and a Global City: The Potential Consequences of Climate Variability and Change, Metro East Coast*, Report for the U.S. Global Change Research Program, Columbia Earth Institute.

Mann, M.E., and P.D. Jones. 2003. Global surface temperatures over the past two millennia. *Geophysical Research Letters*, 30(15):1820-3.

McMahon, T.A., R.J. Nathan, B.L. Finlayson, and A.T. Haines. 1989. Reservoir System Performance and Climatic Change. In *Proceedings of the National Workshop on Planning and Management of Water Resource Systems: Risk and Reliability*, G.C. Dandy and A.R. Simpson (eds.). Canberra, Australia: Australian Government Publishing Service.

Meehl, G.A., W.M. Washington, W.D. Collins, J.M. Arblaster, A. Hu, L. E. Buja, W.G. Strand, and H. Teng. 2005. How Much More Global Warming and Sea Level Rise? *Science*, 307:1769-1772.

Meko, D.M., M.D. Therrell, C.H. Baisan, and M.K. Hughes. 2001a. *Sacramento River Annual Flow Reconstruction*. International Tree-Ring Data Bank. IGBP PAGES/ World Data Center for Paleoclimatology Data Contribution Series #2001-081. Boulder, CO: NOAA/NGDC Paleoclimatology Program.

Meko, D.M., M.D. Therrell, C.H. Baisan, and M.K. Hughes. 2001b. Sacramento River flow reconstructed to A.D. 869 from tree rings. *J. of the American Water Resources Association*, 37 (4): 1029-1040.

Mimikou, M., P.S. Hadjisavva, Y.S. Kouvopolous, and H. Afrateos. 1991. Regional Climate Change Impacts: II. Impacts on Water Management Works. *Hydrological Sciences Journal*, 36:259–270.

Moberg, A., D.M. Sonechkin, K. Holmgren, N.M. Datsenko, and W. Karlén. 2005. Highly variable Northern Hemisphere temperatures reconstructed from low- and high-resolution proxy data. *Nature* 433, 613–617 (10 February).

Mote, P.W., E.A. Parson, A.F. Hamlet, W.S. Keeton, D. Lettenmaier, N. Mantua, E.L. Miles, D.W. Peterson, D.L. Peterson, R. Slaughter, and A.K. Snover. 2003. Preparing for Climatic Change: The Water, Salmon, and Forests of the Pacific Northwest. *Climatic Change*, 61(1-2):45–88.

Mote, P.W. 2004. *The West's Snow Resources in a Changing Climate*. Testimony before the U.S. Senate Committee on Commerce, Science, and Transportation, May 6, 2004.

Nash, L.L., and P.H. Gleick. 1993. *The Colorado River Basin and Climatic Change: The Sensitivity of Streamflow and Water Supply to Variations in Temperature and Precipitation*. Report, US Environmental Protection Agency, Office of Policy, Planning and Evaluation, Climate Change Division, EPA 230-R-93-009. Oakland, CA: Pacific Institute for Studies in Development, Environment, and Security.

Nash, L.L., and P.H. Gleick. 1991. The Sensitivity of Streamflow in the Colorado Basin to Climatic Changes. *Journal of Hydrology*, 125:221–241.

Nemec J., and J. Schaake. 1982. Sensitivity of Water Resource Systems to Climate Variation. Hydrological Sciences, 27(3):327–343.

Palmer, R.N., and M. Hahn. 2002. The Impacts of Climate Change on Portland's Water Supply: An Investigation of Potential Hydrologic and Management Impacts on the Bull Run System. Seattle: Department of Civil and Environmental Engineering, University of Washington.

Pan, Z., R.W. Arritt, E.S. Takle, W.J. Gutowski, C.J. Anderson, and M. Segal. 2004. Altered Hydrologic Feedback in a Warming Climate Introduces a "Warming Hole." *Geophysical Res. Letters*, 31:L17109.

Pelto, M.S. 1993. Changes in Water Supply in Alpine Regions Due to Glacier Retreat. In The World at Risk: *Natural Hazards and Climate Change*. Conference Proceedings. New York: American Institute of Physics.

Pomeroy, J.W., and D.M. Gray. 1995. *Snowcover - Accumulation, Relocation and Management*. Saskatoon, Canada: National Hydrology Research Institute Science Report No. 7

Richardson, C.W.. 1981. Stochastic Simulation of Daily Precipitation, Temperature, and Solar Radiation. *Water Resources Research*, 17:182–190.

Rosgen, David. 1996. Applied River Morphology. Wildland Hydrology, Pagosa Springs, CO. 343 pp.

Shiklomanov, I.A. 1998. *World Water Resources: A New Appraisal and Assessment for the 21st Century*. Paris, France: International Hydrological Programme, UNESCO.

Solley, W.B., R.S. Pierce, and H.A. Perlman. 1998. *Estimated Use of Water in the United States in 1995*, U.S. Geological Survey Circular 1200.

Stewart, I., D.R. Cayan, and M.D. Dettinger. 2004. Changes in Snowmelt Runoff Timing in Western North America under a "Business as Usual" Climate Change Scenario. *Climatic Change*, 62:217–232.

Stewart, I.T., D.R. Cayan, and M.D. Dettinger. 2005. Changes toward earlier streamflow timing across Western North America. *Journal of Climate*, 18: 1136-1155.

Stockton, C. W. and G. C.Jacoby, 1976. Long-Term Surface Water Supply and Streamflow Levels in the Upper Colorado River Basin, Lake Powell Research Project Bulletin No 18., Inst. of Geophysics and Planetary Physics, University of California, Los Angeles, 70 pp.

Stockton, C., and G.C. Jacoby. 2004. *Colorado River Basin Streamflow Reconstructions*. IGBP PAGES/World Data Center for Paleoclimatology Data Contribution Series #2004-017. NOAA/NGDC Paleoclimatology Program, Boulder CO, USA.

Subak, S. 2000. Climate Change Adaptation in the U.K. Water Industry: Managers' Perceptions of Past Variability and Future Scenarios. *Water Resources Management*, 14:137–156.

Trenberth, K.E., A. Dai, R.M. Rasmussen, and D.B. Parsons. 2003. The Changing Character of Precipitation. *Bulletin of the American Meteorological Society*, 84(9): 1205–1217.

Trenberth, K.E., K. Miller, L. Mearns, and S. Rhodes. 2000. Weather Associated with Changing Climate and Consequences for Human Activities. Boulder, CO: University Corporation for Atmospheric Research.

U.S. Geological Survey. 2000. *Sea Level and Climate,* USGS Fact Sheet 002-00. Available at: http://pubs.usgs.gov/factsheet/fs2-00/.

Washington, W. 1996. An Overview of Climate Modeling. In *Final Report: An Institute on the Economics of the Climate Resource*, K.A. Miller and R.K. Parkin (eds.). Boulder, CO: University Corporation for Atmospheric Research.

Wigley, T.M.L. 2004. Modeling climate change under no-policy and policy emission pathways. In *The Benefits of Climate Change Policies: Analytical and Framework Issues*. Paris: OECD Publications, 221–248.

Wigley, T.M.L. 2005. The Climate Change Commitment. *Science*, 307: 1766-1769.

Wilby, R.L., S.P. Charles, E. Zorita, B. Timbal, P. Whetton, and L.O. Mearns. 2004. *Guidelines for Use of Climate Scenarios Developed from Statistical Downscaling Methods*. Report prepared for the IPCC Task Group on Data and Scenario Support for Impacts and Climate Analysis (TGICA). http://ipcc-ddc.cru.uea.ac.uk/guidelines/dgm_no2_v1_09_2004.pdf (accessed May 20, 2005).

Wilks, D.S. 1992. Adapting Stochastic Weather Generation Algorithms for Climate Change Studies. *Climatic Change*, 22:67–84.

WMO (World Meteorological Organization). 2003: *Climate: Into the 21st Century*, W. Burroughs (ed.). Cambridge, UK: Cambridge University Press

Woodhouse, C.A. and Overpeck, J.T. 1998. 2000 years of drought variability in the Central United States. *Bulletin of the American Meteorological Society*, 79, 2693-2714.